Designing and Conducting a Forest
Inventory - case: 9th National Forest
Inventory of Finland

Managing Forest Ecosystems

Volume 21

Series Editors:

Klaus von Gadow
*Georg-August-University,
Göttingen, Germany*

Timo Pukkala
*University of Joensuu,
Joensuu, Finland*

and

Margarida Tomé
*Instituto Superior de Agronomía,
Lisbon, Portugal*

Aims & Scope:

Well-managed forests and woodlands are a renewable resource, producing essential raw material with minimum waste and energy use. Rich in habitat and species diversity, forests may contribute to increased ecosystem stability. They can absorb the effects of unwanted deposition and other disturbances and protect neighbouring ecosystems by maintaining stable nutrient and energy cycles and by preventing soil degradation and erosion. They provide much-needed recreation and their continued existence contributes to stabilizing rural communities.

Forests are managed for timber production and species, habitat and process conservation. A subtle shift from *multiple-use management* to *ecosystems management* is being observed and the new ecological perspective of *multi-functional forest management* is based on the principles of ecosystem diversity, stability and elasticity, and the dynamic equilibrium of primary and secondary production.

Making full use of new technology is one of the challenges facing forest management today. Resource information must be obtained with a limited budget. This requires better timing of resource assessment activities and improved use of multiple data sources. Sound ecosystems management, like any other management activity, relies on effective forecasting and operational control.

The aim of the book series ***Managing Forest Ecosystems*** is to present state-of-the-art research results relating to the practice of forest management. Contributions are solicited from prominent authors. Each reference book, monograph or proceedings volume will be focused to deal with a specific context. Typical issues of the series are: resource assessment techniques, evaluating sustainability for even-aged and uneven-aged forests, multi-objective management, predicting forest development, optimizing forest management, biodiversity management and monitoring, risk assessment and economic analysis.

For further volumes:
http://www.springer.com/series/6247

Erkki Tomppo • Juha Heikkinen
Helena M. Henttonen • Antti Ihalainen
Matti Katila • Helena Mäkelä
Tarja Tuomainen • Nina Vainikainen

Designing and Conducting a Forest Inventory - case: 9th National Forest Inventory of Finland

Prof. Erkki Tomppo
The Finnish Forest Research Institute
Jokiniemenkuja 1
01301 Vantaa
Finland
erkki.tomppo@metla.fi

Helena M. Henttonen
The Finnish Forest Research Institute
National Forest Inventory/Vantaa
Jokiniemenkuja 1
01301 Vantaa
Finland
helena.henttonen@metla.fi

Dr. Matti Katila
The Finnish Forest Research Institute
Jokiniemenkuja 1
01301 Vantaa
Finland
matti.katila@metla.fi

Tarja Tuomainen
The Finnish Forest Research Institute
Greenhouse Gas
Reporting/Vantaa
Jokiniemenkuja 1
01301 Vantaa
Finland
tarja.tuomainen@metla.fi

Juha Heikkinen
The Finnish Forest Research Institute
National Forest Inventory/Vantaa
Jokiniemenkuja 1
01301 Vantaa
Finland
juka.heikkinen@metla.fi

Antti Ihalainen
The Finnish Forest Research Institute
National Forest Inventory/Vantaa
Jokiniemenkuja 1
01301 Vantaa
Finland
antti.ihalainen@metla.fi

Helena Mäkelä
The Finnish Forest Research Institute
National Forest Inventory/Vantaa
Jokiniemenkuja 1
01301 Vantaa
Finland
helena.makela@metla.fi

Nina Vainikainen
The Finnish Forest Research Institute
National Forest Inventory/Vantaa
Jokiniemenkuja 1
01301 Vantaa
Finland
nina.vainikainen@gmail.com

ISSN 1568-1319
ISBN 978-94-007-1651-3   e-ISBN 978-94-007-1652-0
DOI 10.1007/978-94-007-1652-0
Springer Dordrecht Heidelberg London New York

Library of Congress Control Number: 2011933556

© Springer Science+Business Media B.V. 2011
No part of this work may be reproduced, stored in a retrieval system, or transmitted in any form or by any means, electronic, mechanical, photocopying, microfilming, recording or otherwise, without written permission from the Publisher, with the exception of any material supplied specifically for the purpose of being entered and executed on a computer system, for exclusive use by the purchaser of the work.

Printed on acid-free paper

Springer is part of Springer Science+Business Media (www.springer.com)

Preface

National Forest Inventories in Finland have evolved gradually over a period of one hundred years, first with a trial inventory in the 1910s, and since the 1920s, with operative inventories. The sampling design and estimation methods have been continuously revised to correspond with inventory techniques and the available infrastructure. The content, as well as the collected data and variables employed, are constantly adapted to the users' needs. Co-operation with the inventories of other countries, particularly with those of the Nordic countries, has supported these developments. Assessing the wood production potential was the key aim of the early inventories, and has remained important in the current inventories. Forest health and damage assessments were added in the 1970s and 1980s. The assessment of biodiversity became an important part of the inventory in the 1990s with the UNCBD agreements. National inventories worldwide are vital information sources for greenhouse gas reporting for the UNFCCC and its Kyoto Protocol.

This book demonstrates in detail all phases of the 9th National Forest Inventory of Finland (1996–2003): the planning of the sampling design, measurements, estimation methods and results. The inventory knowledge accumulated during almost one hundred years is consolidated in the book. The purpose of the numerous examples of results is to demonstrate the diversity of the estimates and content of a national forest inventory. The authors hope that the book will help in designing and conducting any large area forest inventory.

Helsinki December 2010
Erkki Tomppo, Juha Heikkinen,
Helena M. Henttonen, Antti Ihalainen,
Matti Katila, Helena Mäkelä,
Tarja Tuomainen, Nina Vainikainen

Acknowledgements

The main results of this book are based on the 9th National Forest Inventory of Finland and have required knowledge and input from a number of people. Erkki Tomppo, Kari T. Korhonen and Tarja Tuomainen have been in charge of the fieldwork, 2, 3 and 3 years, respectively. Arto Ahola's role was crucial in training the field crews and in conducting the fieldwork, as well as programming the first version of field computer and preparing and checking the data. The following persons worked as the field crew leaders: Jouni Kulju, Juhani Kumpuniemi, Rauno Salo, Pertti Virtanen, Jouni Peräsaari and Tuomo Saastamoinen 7–8 seasons; Arto Ahola, Anssi Korsström, Mikko Korhonen, Jouni Kunnari, Markku Pernu, Juha Pulli, Jarmo Tuomainen, Mari Honkonen, Esa Kinnunen, Juhani Moilanen and Kati Tammela 2–4 seasons; and Mikko Heikkinen, Timo Hongisto, Olli-Pekka Jalonen, Matti Katila, Kimmo Kivinen, Kari T. Korhonen, Jouko Kytölä, Pertti Lepolahti, Juha Leppälä, Tarja Manninen, Nina Mäkinen, Sampo Parviainen, Taina Sairio, Pasi Sarriolehto, Arto Sorri, Erika Tammilehto and Jari Varjo one season. Alpo Aarnio had a central role in implementing the method for volume predictions for the tally trees. Mikael Strandström implemented the second version of the field computer and data transfer. Kai Mäkisara designed and implemented the GPS system and participated in the field computer system design. Major technical assistance was provided by Karipekka Byman, Pekka Oksman, Ulla Suhonen and Kati Tammela.

Numerous units and individuals at Metla participated in planning of the field measurements. To mention only a few of them, Tiina Tonteri had a central role in the planning the key biotope measurements, as well as deadwood measurements together with Juha Siitonen and Reijo Penttilä. Eino Mälkönen coordinated the revision of forest health measurements, Pekka Tamminen the mineral soil measurements and Erkki Ahti together with Seppo Kaunisto the peatland soils measurements. Antti Reinikainen managed mineral soil site fertility classification. Markku Siitonen and his team helped to select the variables to be employed in calculating the alternative cutting scenarios. Seppo Nevalainen, Hannu Nousiainen, Antti Reinikainen and Tiina Tonteri actively participated in the training of field crews, in addition to NFI staff.

Forest experts from the following organisations either actively participated in the training of the field crews or provided assistance in conducting the field work or both: Forestry Centres, Tapio consulting services, Forest Industry Companies, Metsähallitus, Finnish Tax Administration, the Central Union of Agricultural Producers and Forest Owners, the Finnish Border Guard, The Gulf of Finland Cost Guard, West Finland Coast Guard and Provincial Administration Åland.

For this publication, Kai Mäkisara, Metla, provided his help in writing paragraphs concerning the GPS devices and Saara Ilvesniemi and Mia Landauer gave technical assistance. The English language was edited by Dr. Ashley Selby. Sari Elomaa was responsible for the demanding typesetting. We are deeply indebted to these and all the individuals whose support has made this book possible. We also express our sincere thanks to the Editorial Board of the Springer's series Managing Forest Ecosystems-series for accepting this book for publication, as well as to Dr. Catherine Cotton, Publishing Editor, and Dr. Valeria Rinaudo, Associate Editor, at Springer for their constant guidance and assistance during all the phases of the production of the book.

Contents

1	**Introduction**	1
1.1	Climatic Conditions and Forests of Finland	2
1.2	Early Attempts to Assess Forest Resources	2
1.3	The Development of the National Forest Inventories in Finland	5
1.4	The Use of the Forest Inventory Results in Forest Policy	8
1.5	The Use of National Forest Inventory Data in the UNFCCC and the Kyoto Protocol Reporting	10
1.6	The Role of National Forest Inventory in Assessing the Status of Biodiversity	11
1.7	The Content and Structure of the Book, Further Results of NFI9	12
References		13
2	**Design and Measurements**	17
2.1	Field Sampling Design	17
	2.1.1 Sampling Simulation	20
	2.1.2 South Finland	20
	2.1.3 North Finland (Excluding North Lapland)	23
	2.1.4 North Lapland	23
2.2	Assessment Units	25
	2.2.1 Angle Count Plots	25
	2.2.2 Stands	26
	2.2.3 Other Assessment Units	26
2.3	Locating the Field Plots	27
2.4	Administrative Information	28
2.5	Land-Use and Classification of Forestry Land	30
2.6	Site Variables	31
2.7	Soil Variables	33
2.8	Drainage Situation	35
2.9	Taxation Class	35

	2.10	Retention Trees to Maintain Biodiversity of Forests	36
	2.11	Description of the Growing Stock of the Stand	37
	2.12	Damages	42
	2.13	Silvicultural Quality of Stand	44
	2.14	Accomplished and Proposed Measures	45
	2.15	Key Habitat Characteristics	46
	2.16	Tally Tree Measurements	49
	2.17	Epiphytic Lichens	55
	2.18	Keystone Tree Species	55
	2.19	All Tree Species	56
	2.20	Dead Wood Measurements	56
	2.21	Equipment for Measurements	58
	2.22	A Correction to the Height Measurements of Year 2001	60
		2.22.1 The Height Correction Models for the Sample Trees not Re-measured	61
		2.22.2 Models for Correcting the Height Increments	62
	2.23	Training and Quality Assurance	63
	2.24	The Workload and Costs	64
References			65
3	**Estimation Methods**		**69**
	3.1	Estimation of Areas	70
	3.2	Estimation of the Current Growing Stock	71
		3.2.1 Mean Values per Area Unit	72
		3.2.2 Mean Diameters	74
		3.2.3 Predicting Sample Tree Form Factors, Volumes and Timber Assortment Proportions	75
		3.2.4 Predicting Form Heights for Tally Trees	78
	3.3	Estimation of Volume Increment	80
		3.3.1 Increment of a Sample Tree	81
		3.3.2 Increment of Survivor Trees	82
		3.3.3 Increment of Drain	83
		3.3.4 Total Increment	84
	3.4	Estimation of the Volume of Dead Wood	85
	3.5	Assessment of Sampling Error	85
		3.5.1 Sampling Error of Ratio Estimators	86
		3.5.2 Sampling Error of Total Volumes and Aggregates	88
	3.6	Thematic Maps	89
References			90
4	**Results**		**93**
	4.1	The Areas of Land-Use Classes and Their Development	93
		4.1.1 Forestry Land	93
		4.1.2 Forest Land	96
		4.1.3 Land Classes Based on FAO Definitions	97
		4.1.4 Land Use-Changes Based on the Observations on the Plot	98
		4.1.5 Ownership Information	99

Contents

4.2	Restrictions on Forestry and Area Available for Wood Production	100
4.3	Soil Classification and the Areas of Site Fertility Classes on Mineral Soils	101
4.4	Peatlands and Their Site Classes	104
	4.4.1 Peatland Area and Its Changes	104
	4.4.2 Land Classes of Peatlands	106
	4.4.3 Drainage Situation of Peatlands	107
	4.4.4 Principal Site Classes and Site Fertility Classes on Peatland Soils	109
	4.4.5 The Thickness of the Peat Layer	111
4.5	Tree Species Dominance and Composition	113
	4.5.1 The Dominant Tree Species	113
	4.5.2 Tree Species Dominance by Site Fertility Classes	118
	4.5.3 Tree Species Mixtures	119
4.6	Age and Development Classes	121
	4.6.1 The Age Distributions of Stands and Their Changes	121
	4.6.2 The Development Classes of Stands and Their Changes	124
4.7	Growing Stock	126
	4.7.1 Mean Volume Estimates by Tree Species	126
	4.7.2 Total Growing Stock Estimates	132
	4.7.3 Volume Estimates of Saw-Timber	135
4.8	Volume Increment	136
	4.8.1 Increment Estimates	136
	4.8.2 Uncertainties in Increment Estimates and Comparisons with Estimates from Earlier Inventories	136
	4.8.3 Forest Balance	140
	4.8.4 Changes in Annual Volume Increment Estimates since the 1950s	143
4.9	Protected Areas	148
4.10	Forest Damage	150
4.11	Silvicultural Quality of Forests	154
	4.11.1 Silvicultural Quality	154
	4.11.2 Methods and Success of Regeneration	156
4.12	Management Activities	157
	4.12.1 Accomplished and Proposed Cuttings	157
	4.12.2 Accomplished and Proposed Silvicultural Measures	159
	4.12.3 Drainage Operations	160
4.13	Biodiversity Indicators	162
	4.13.1 Biodiversity Measurements in NFI	162
	4.13.2 Key Habitats	163
	4.13.3 Dead Wood	165

		4.13.4	Keystone Tree Species	171
		4.13.5	Retention Trees	172
References				174
5	**Discussion**			179
	5.1	The Development of Volume and Increment: New Estimates from NFI9		179
	5.2	Estimation and Error Estimation		180
	5.3	Comparisons of the NFI8 and NFI9 Designs		181
	5.4	Some Experiences of the New Measurements		182
	5.5	Experiences with the New Measurement Devices		182
	5.6	Changes in the Design for NFI10		183
References				184
Appendix				185
Index				261

Chapter 1
Introduction

Reliable information on forest resources and the status of forests at the national level is required for various purposes, such as when making strategic decisions, formulating regulations and recommendations for forest management, aimed at ensuring the availability of sufficient supply of timber for the forest industries. Information is also needed by the companies for the strategic planning of investments. It is also essential in planning forest protection and for maintaining biodiversity. Global agreements, such as the Kyoto Protocol, increasingly rely on national forest statistics in the negotiation phase, in the monitoring of their compliance and in reporting.

The assessment of the forests of a nation is a challenging task in many respects. First, the number of trees is enormous, e.g. in Finland roughly 80 billion with a height of more than 1.3 m. It is thus impossible to measure all individual trees in a realistic time. Another aspect is that information is needed not only about trees, but also about soil, the site, as well as e.g., accomplished and necessary silvicultural and cutting activities. A solution in early forest inventories was to assess the tree resources ocularly for some areal units throughout the entire forest area of interest. One problem, in addition to the subjectivity of the method, was the lack of methods for assessing the errors of the estimates, as well as the amount of time and considerable resources required. An elegant solution to such problems was sampling with precise measurements, and it is still valid. This approach also makes it possible to derive error estimates based on the statistical theory.

The aim of this book is to demonstrate in detail all phases of a modern sampling-based national forest inventory: planning, measurements, estimation and results. This is made as concrete as possible by focusing on just one example, the 9th National Forest Inventory of Finland. The role of forests has long been exceptionally important for Finnish society and economy, and the history of the national inventories is among the longest in the world. As a result of this considerable experience and continuously active research, the Finnish inventory can serve as a good example.

Excellent textbooks on forest inventory methodologies have been written. These include books by Loetsch and Haller (1964), Loetsch et al. (1973), Schreuder et al. (1993), Mandallaz (2008) and Gregoire and Valentine (2008). The book edited by Kangas and Maltamo (2006) presents some methodological articles with country cases while the edited book by Tomppo et al. (2010a) presents many examples of national forest inventories (NFI), but does not go in such details as this present book. This volume complements an earlier book partly by the same authors, published in the same series and demonstrating the method called *multi-source forest inventory* (MS-NFI) (Sect. 1.7) (Tomppo et al. 2008).

1.1 Climatic Conditions and Forests of Finland

Finland is located roughly between the latitudes 60° and 70° north and is therefore one of the most northern forested countries in the world. The length of the growing season ranges from 190 days in the most southern Finland to about 60 days in the North and the effective temperature sum from 1,400 to 200 day degrees (Fig. 1.1a, b). The favourable climate is largely caused by the location of the country in relation to the zone of westerly winds and its close proximity to the Atlantic Ocean and Baltic Sea that cause the winters to be milder than they would be otherwise. The continental land mass to the east makes the summers warm over most of the area (Tuhkanen 1980).

In NFI9, forests accounted for 75% of the land area. However, southern and northern parts of the country differ significantly with respect the growth conditions and so forest resource estimates are often presented separately for those regions, in addition to the other regional results.

1.2 Early Attempts to Assess Forest Resources

Slash-and-burn agriculture, the production of wood to tar, widespread household uses of timber, the making of charcoal for iron smelting, and from the second half of nineteenth century, an increasing forest industry exhausted Finnish forest resources. According to pre-inventory documents, the forest resources of Finland were at their lowest level when slash-and-burn cultivation ended, at the turn of the twentieth century (Myllyntaus and Mattila 2002). The rapid growth of forest industry in the 1870s and the extension of cuttings deeper into wilderness forests stimulated the interest in forest resources.

The first attempts to assess the forest area and forest resources of Finland were made during the mid nineteenth century. For example, by C. W. Gyldén (1853) assessed the area of "land growing forest" to be 59.2% of the total land area. The lack of quantitative information in those early assessments was completed with verbal descriptions. One such assessment on the state of Finnish forests was made

1.2 Early Attempts to Assess Forest Resources

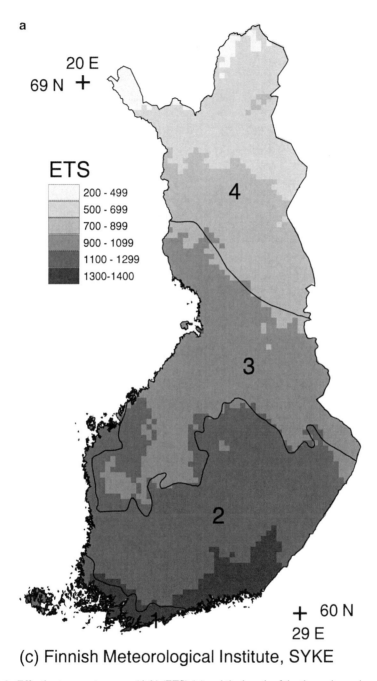

Fig. 1.1 Effective temperature sum (d.d.) (ETS) (**a**) and the length of the thermal growing season (days) (**b**) with Forest vegetation zones: *1* Hemiboreal, *2* Southern boreal, *3* Middle boreal, *4* Northern boreal

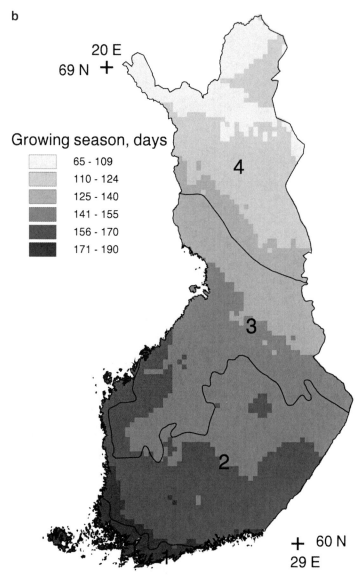

Fig. 1.1 (continued)

by von Berg (1858), after a tour through Finland. This assessment was attached to a committee report that considered the exhaustion of forest resources.

An attempt to systematically and reliably describe the forests in the whole country began in 1880. The studies by A. G. Blomqvist, the Director of the Evo Forestry

Institute, concerning the international position of Finland as a forest producing country, acted as the chief impulse (Blomqvist 1881; Ilvessalo 1927). The assessment made by Hannikainen (1896) is considered to have been more precise than the previous ones (Ilvessalo 1927). However, these early attempts did not lead to any significant results due to, e.g., the economical crisis in the end of the nineteenth century. In 1907, Professor A. K. Cajander made a proposal for the Finnish Forestry Association, Tapio. The intention was to make a trial survey in two small municipalities to investigate, among other things, what distance between the survey lines could be used in different areas without affecting the accuracy of the lower limit allowed (Ilvessalo 1927). It was planned that two large municipalities should be surveyed first, and then the entire country. Only the first part of the plan was completed, in the municipalities of Sahalahti and Kuhmalahti. The inventory was planned and started by Cajanus (1913) and completed by Ilvessalo (1923).

Trial inventories were also carried out in Norway and Sweden and statistical methods for forest inventory were developed, often in co-operation between the scientists of the three countries. Trial inventories were made, e.g., in Øsfold and Hedmark counties in Norway and in Värmland county in Sweden (Langsæter 1926; Östlind 1925, 1932). Loetsch and Haller (1964, p. 9) give a list of the important development work for statistical forest surveys. In addition, Cajanus (1913) studied the correction of the bias of ocular estimation, which presented an early idea for the error estimation in line survey sampling. The idea was to use "random variation" between the inventory lines as the basis for error estimation after removing the "real variation"; the spatial trend. The spatial trend was modelled and assessed by fitting the trend curve visually to a point set of the line level results when the results (y-axis) are plotted against the line number (x-axis, representing the distances between the lines). This can be considered to be an early version of model-based error estimation, and it was used by Ilvessalo (1923) in the Sahalahti and Kuhmalahti inventory, see also Langsæter (1926) and Heikkinen (2006). Cajanus' approach is, however, vulnerable to subjective trend curve fitting. To improve the method for the first NFI, Ilvessalo turned to a Finnish mathematician Lindeberg (see Sect. 1.3). He presented a method based on the variability of the differenced series in which a separate trend fitting was not required (Lindeberg 1924, 1926). For other early Nordic examples, see e.g., Langsæter (1932) and Östlind (1932).

1.3 The Development of the National Forest Inventories in Finland

The plan to continue the trial survey of Sahalahti and Kuhmalahti to the national level was abandoned for many reasons. The plan became outdated in many respects with the rapid progress of forest science. Furthermore, the country had gained its independence and a new up-to-date plan was necessary. Nevertheless, the experiences gained from earlier inventories provided beneficial information (Ilvessalo 1927).

The initiative for the first national forest inventory was provided by the Tax Reform Committee in 1921 that suggested research to investigate the possibilities for taxing the forest yield. A line passing from Loppi in Southern Finland to Pielisjärvi (currently Lieksa) in North Karelia with a length of 315.855 km, was measured in the same year, but the information was found to be inadequate. A new suggestion brought a result and a new start was made in summer 1922. The field measurements were completed in 1924 (by measuring additional lines in the county of Uusimaa in South Finland).

This work was lead by Professor Yrjö Ilvessalo. The areas of site and stand classes, as well as growing stock characteristics were estimated using the line survey, and plot measurements were employed to calibrate the ocular assessments on the volumes and increment on stands. Lines were oriented from South-West to North-East at distances of 26 km in most parts of the country, but at 13 km in Uusimaa (in South Finland) and 10 km in Åland. For each land figure and forest stand touched by the survey line, land use, soil, site and growing stock characteristics were assessed ocularly. The assessment concerned the forest stand and were, e.g., for volumes, averages per hectare. To obtain precise growing stock and increment estimates, sample plot measurements were carried out in line strips with a 10 m wide. The plot length was 50 m and the interval between two plots on a line was 2 km. More precisely, on the basis of the basic rule, the trees were measured on the last 50 m of each line part of 2 km. However, the plot had to be shifted in such a way that it was entirely inside of one forest stand. The trees thicker than 20 cm at a breast height were callipered also on the first line part of each 2 km line section. The plot of the thick trees was not shifted. Regression lines were estimated for volume and volume increment estimates measured on the plots as a function of the ocular assessment by field crews, and in South part of the country separately for low and high volumes. These regression lines were used for correcting the biases of the ocular estimates. The breast height diameter $(d_{1.3})$ was measured at the height of 1.3 m above the ground level.

The forest inventories were established almost at the same time as in Finland, in Norway, in 1919, and in Sweden, in 1923. It should be noted that the method of the Finnish inventory deviated from those in Norway and Sweden, where trees were callipered along the entire strip. Strip surveys require a large amount of work and the method is vulnerable to errors. Errors caused by the boundary trees are discussed, e.g., in Ilvessalo (1927): " ...in spite of constant vigilance, there could not have been any guarantee of a very careful checking of the numerous trees occurring close to the outer boundary of the strip" (Ilvessalo 1927, p. 342). These early findings in Nordic countries still have relevance when selecting the field plot size and shape in any country. The question of field plot size and shape is topical worldwide in countries where forest inventories are established for the first time, or are being revised.

The main results of the survey were published as already in 1924 (Ilvessalo 1924) and presented at the World's Forestry Congress in Rome 1926. The publishing of the complete results in 1927 was a historical moment. Knowledge of forest resources based on a national forest inventory with a sound statistical framework was available

1.3 The Development of the National Forest Inventories in Finland

for the first time. This at least partly influenced the establishment of a sample based inventory in USA when President of the USA, Calvin Coolidge invited Professor Yrjö Ilvessalo to discuss about the inventory (Labau et al. 2007; Tomppo et al. 2010b). The earliest large-area, sample-based inventories date to the passage of U.S. federal legislation in 1928.

The first inventory in Finland (NFI1) showed the forest resources to be more plentiful than assumed. However, the rapid growth of the pulp industry raised questions about the sufficiency of forest resources. The second inventory (NFI2) was carried out during the years 1936–1938. The line directions were the same as in the first inventory, but the distance between the lines in southern Finland and in the southern part of North Finland was 13 km, 26 km elsewhere in North Finland and 10 km in Åland. The sample plot distance along a line was 1,000 m, and the size of the circular plot 0.1 ha, except for trees with breast height diameter ($d_{1.3}$) less than 5 cm 0.01 ha, and in advanced seedling stands and in other stands with trees of same size 0.05 ha and in dense young seedling stands 0.01 ha. The minimum diameter for broadleaved trees was 2.5 cm. The diameter $d_{1.3}$ was measured differently than in NFI1: 1.3 m above the upper root collar that was assumed to hamper felling. The ocular assessments of volumes and volume increments from the land figures touching the lines were again corrected using regression lines (Ilvessalo 1942). The results showed that the forest resources had survived the demands of the forest industries.

The method in the NFI3, (1951–1953) was almost the same as in the second one. The distance between the lines was 13 km for most parts of the country; but, in the northern parts where forestry is more extensive, it was 16 km, and in the northernmost part of Lapland 20 km. In some of the southern districts, the distance between the lines was 6.5 km. Sample plots of 0.1 ha were measured at 1 km intervals; and halfway between them, sample plots of the same size were established for trees with a minimum $d_{1.3}$ of 18 cm. Aerial survey was used as a source of an auxiliary information in areas with difficult access (Ilvessalo 1956). The forest inventory results published in 1956 showed the forest resources to be almost as large as before the war, in spite in the 10% decrease of the forest area resulting from land losses in 1945 after the war. The main reason was the diminished cuttings during the war. Ground vegetation observations, made by a biologist in each field crew (16–17 crews) were an additional feature of the NFI3. The results concerning the abundance and occurrence of various plant species were published by Kujala (1964). More recent results were published by Reinikainen et al. (2000) and compared to those by Kujala (1964).

Expansions of forest industries in the 1950s and the export of timber gave rise to a new threat for the sufficiency of forests. The fourth inventory was carried out during the years 1960–1961. It was originally intended to be an intermediate inventory in the southern part of the country. During the inventory, a new method should have been developed. However, the inventory was expanded to the entire country (Ilvessalo 1962, 1963). The distance between the lines was approximately the same as in the first inventory. Land classes and tree species dominance were assessed on the stands touching the lines. The volume and increment estimates were based on the measurements along the lines and sample plots, without visual estimation. The plot interval on the lines was lower than in the earlier inventories: 250 m, having a size 0.03 ha of the circular plot for trees with $d_{1.3}$ more than 10 cm, and 100 m² for smaller trees.

A pilot inventory in 1963 confirmed the feasibility of the new sampling design. Measurements were carried out on temporary sample plots, and trees were selected using angle count sampling (Bitterlich sampling). The continuous lines were broken into either rectangles (Southern Finland) or half rectangles (Northern Finland). Some stand level variables were assessed from stands crossed by the lines and the line lengths were recorded (Kuusela and Salminen 1969). The inventory organisation became permanent. The fifth NFI moved yearly from south to north by forestry board districts. In North Lapland, aerial photographs were used in addition to the field plots (Poso and Kujala 1971). The method remained almost the same in NFI6 (1971–1976), except that there were fewer stand level measurements on the lines than in NFI5, and the use of aerial photographs and two phase sampling in a large area in Lapland (Poso and Kujala 1977). The stand level measurements along the lines were omitted from NFI7 (1977–1984) but stump plots were added between the tree plots. A combination of field plots and aerial photographs was used throughout Lapland (Mattila 1985).

The eight NFI was completed in 1986–1994. A new feature in the northern part of the country was the permanent sample plots that were re-measured in NFI9. In each cluster, 3 of the 15 plots were marked for re-measuring (Tomppo et al. 2001). In addition to the results based on the field data only, the results were calculated for small areas (municipalities) using what is called a multi-source inventory, i.e., using field data, satellite images and digital map data. The method was adopted during NFI8 in 1990. Some of the temporary field plots, measured in NFI8 in 1986–1989, were re-measured in 1994 in order to obtain up-to-date results, and to apply the multi-source inventory to the southern part of the country (Tomppo et al. 1998b).

Forest resource information can to some extent be updated computationally over a short time period using earlier field measurements, cutting statistics and new growth variation measurements. The first such update concerning the whole country was completed in Spring 1990 employing the data of NFI8 and NFI7. This provided up-to-date estimates of the growing stock in that year and increment estimates for the years 1985–1989.

1.4 The Use of the Forest Inventory Results in Forest Policy

The importance of the forest industries to Finland's national economy has given strategic value to the forest inventory information. In the 1950–1960s, almost 80% of the export income came from forest industry (National Board of Customs 2010). The annual turnover ranged from 19 to 22 billion € during the period 2001–2008. Forest industries have also facilitated other industrial sectors, such as the metallurgical, electrical and information technology industries.

1.4 The Use of the Forest Inventory Results in Forest Policy

The information generated by the Finnish national forest inventory has traditionally been used for large-area forest management planning (e.g. for planning cutting, silviculture and forest improvement regimes at the regional and national levels), for decisions concerning forest industry investments, and as a basis for forest income taxation (until 2006 when the area based forest income taxation was changed to a revenue basis). It has also provided forest resources information for national and international statistics, such as the Forest Resources Assessment (FRA) procedure of the Food and Agriculture Organization of the United Nations (FAO) and the Ministerial Conference on the Protection of Forests in Europe (MCPFE 2007). Furthermore, the NFI produces information on forest health status and damage, biodiversity and carbon pools, as well as changes in these, for the Land Use, Land-Use Change and Forestry (LULUCF) reports of the United Nations Framework Convention on Climate Change (Statistics Finland 2010).

An important driving force, and final impulse for establishing the first national forest inventory in Finland, was the concern about the sufficiency of forest resources for the expanding forest industry, in addition to a need to tax forest land. The taxation requirement, at least partly, arose from a goal to guarantee the sufficiency of timber resources. In addition to the volume and volume increment estimates, important information were the silvicultural status of the forests from the wood production point of view, as well as the volume and quality of saw timber. Also Ilvessalo (1927) pointed out the importance of developing high yield forests and to construct tools appropriate tools to assess the growth potential.

Ilvessalo used an approximate method with the first NFI data and concluded that the mean volume of the forests is much less than the "normal" mean volume. The NFI results and the status of the forests led to a renewal of the Forest Acts. Several new Acts were introduced in 1928: a new Forest Act replacing the one from 1886 (stating, e.g., that forest should not be destroyed and presenting rules concerning the treatment of young forests), an Act concerning funding of forest improvements, and an Act to establish the Forestry Board Districts (current Forest Centres) (Päätös lääninmetsälautakuntien toiminnan lopettamisesta ja metsänhoitolautakuntien toiminnan aloittamisesta 1928). A strict declaration was presented by Appelroth et al. (1948) forbidding selective cuttings. It was based on the studies by Sarvas (1944 and 1950).

Future cutting possibilities have been a part of the inventory results since NFI2. Ilvessalo (1942) presented three different cutting "amounts", (1) the cutting amount estimated on the basis of growth, (2) the cutting amount estimated on the basis of the proportions of age classes, and (3) the silvicultural cutting amount. Since NFI2, the national forest policy and forest industry investments have largely been based on the cutting possibility estimates, sometimes prepared by request.

Several timber production programs have been prepared in Finland since 1960. The main information source for the status of the forests has been the national forest inventories. One of the main goals of the first program, HKLN, was to increase the cutting possibilities of the forests, i.e. by increasing, the volume of the growing stock and the annual increment (Heikurainen et al. 1960). The targets over 40 years were 27% for the volume and 43% the volume increment compared to the levels at the beginning of 1950s. The actual increases observed in the later inventories were 26% and 27% respectively for volume and volume increment. The HKLN program included, e.g., 70,000 ha of peatland drainage annually during the 1960s and 1970s, and approximately 100,000 ha of artificial regeneration of forests per year for the period 1963–1982. The average actual peatland drainage was almost 200,000 ha per year in that period and the regeneration area was also higher than the goal (by nearly 30,000 ha). Several other forecasts of the possibilities of the forests have been made. One was made by Ilvessalo already in 1927, based on NFI1, in which the increment was estimated to be capable of a 36% increase and the volume 31% higher than those found in NFI1. For these goals, the forests would need regular treatment resulting to the growing stock that would correspond the definition of "normal forests". After the HKLN program, Ilvessalo considered possible to double the annual volume increment to 94 million m^3 (Ilvessalo 1961; Nyyssönen 1998).

After the HKLN program, several other timber production programs, or forestry programs, have been put into practise in Finland, e.g., MERA programs I-III (Ervasti et al. 1965, 1970; MERA III 1969) as well as Forest 2000 (Metsä 2000 ohjelma-jaosto 1985) and its update in 1992 (Metsä 2000 ohjelmajan tarkistustoiminkunta 1992). The most recent one is the (National Forestry Program 2010). NFI data have played a central role in all of these programs. In particular, calculations using the MELA system and NFI data were made for Forest 2000 and its up-date. During NFI8, MELA replaced the earlier methods in calculating alternative timber production possibilities. It optimizes the future forest revenue with the given timber prices, rate of interest, predicted future growth, the predictions based on NFI data, as well as the growing stock and its structure in the beginning of the calculation period (MELA ja metsälaskelmat 2010). Several assumptions are made in the system. Some of which are not necessarily fulfilled in practice, for example, the forest owners are assumed to follow good silvicultural practices, recommendations and rules given in the forest act and guidelines, as well as employ optimal timing and order of the cuttings.

1.5 The Use of National Forest Inventory Data in the UNFCCC and the Kyoto Protocol Reporting

The Finnish NFI has been the main data provider for greenhouse gas reporting for the LULUCF sector to the UNFCCC since the mid-1990s, and it also plays a central role in reporting LULUCF activities under Article 3, paragraphs 3 and 4, of the Kyoto Protocol. The definitions of Finnish land-use categories in the reports are consistent with the Good Practice Guidance for LULUCF (IPCC 2003; Statistics Finland 2010). Areas are estimated from the NFI data, but exclude peat production areas that are obtained from other statistics. IPCC land-use categories are created on the basis of national land classes and other land-use related variables assessed in the NFI. The data from NFI7 onwards are employed in estimating time-series of areas for land-use categories from 1990 to the present.

Additionally, the NFI supplies information on soils and it is used to divide forest land into mineral and organic soil types, as well as into undrained and drained lands. The soil information has been used to estimate carbon stock changes on mineral soils and the emissions caused by the drainage of peatlands. Area estimates and tree level data have been employed to estimate tree biomass stocks and biomass increment. The biomass conversion factors for harvest removals and natural losses have also been computed from NFI6 to the present using the permanent sample plots. A comprehensive description of the system is given in the National Inventory Report (Statistics Finland 2010).

New land-use related variables and classifications are included in NFI10 (2004–2008) and NFI11 (2009–2013) to provide adequate information about land-use changes necessary for greenhouse gas reporting. These additions are expected to give a more accurate basis for reporting emissions and removals from afforestation, reforestation and deforestation activities under the Kyoto Protocol.

1.6 The Role of National Forest Inventory in Assessing the Status of Biodiversity

The conventional NFI variables, for example, stand and tree age, tree species composition, diameter distribution of the volume of the growing stock and the number of trees by species and diameter classes, as well as soil and site variables can be used for assessing the status of biodiversity of forests. Site variables in particular, and the calculated results, were used when assessing the extent of potential key habitats and habitats to be included in and protected by the new Finnish Forest of 1996 (Sect. 2.15). These key habitats and their variants were measured in NFI9, together with some additional variable groups, in order to assess the status of forest biodiversity. The new variables are discussed in Sects. 2.15, and 2.18–2.20 and the results in Sect. 4.13. The NFI data are a key source of information when assessing the need and possibilities for further forest protection. The data were also used to estimate the extent of different threatened habitat types in Finland (Raunio et al. 2008). Lehtomäki et al. (2009) used MS-NFI data (Sect. 1.7) to develop methods for prioritizing and establishing new protected areas when the total area to be protected is given.

The NFI field plots grid has been used as a sampling framework to collect data for special assessments and information needs. An example of such an assessment is the amount and biomass of ground lichen species suitable for reindeer grazing. The data were collected and the assessments made in connection with NFI7, NFI8 and NFI9 in the reindeer husbandry region in North Finland and continued in NFI10 (Mattila 1996; Tomppo et al 2001). Räsänen et al. (1985) studied the status of seedling stands in southern Finland with the help of NFI7 data and additional variables collected from NFI field plot stands for this purpose.

NFI data has provided excellent opportunities for several types of research. Järvinen et al. (1977) analysed the landscape structure from the point of view of bird habitats.

Pakkala et al. (2002) and Kangas et al. (2010) made similar studies with multi-source NFI data. NFI data can also been used to derive and calibrate tree and landscape models. Henttonen et al. (2009) studied the progress of the radial increment in Scots pine and Norway spruce during the growing season and the variation of the length of the radial-increment period among years using data from NFI7–NFI10.

1.7 The Content and Structure of the Book, Further Results of NFI9

The book presents a detailed picture of a large-area forest inventory, the field sampling design and methods to plan the design, the plot configuration, field measurements, estimation methods, as well as the inventory results. The sampling design, its development and field measurements with the definitions of the variables are described in Chap. 2, the estimation methods in Chap. 3 and examples of the results in Chap. 4. The numerous results are presented as Appendix tables. The purpose of the results is to demonstrate what type of the estimates can be made from the measurements of a large area inventory. Although the result section is fairly large, even more diverse results could be derived from a national forest inventory. Examples of these additional estimates not presented in this book include the carbon pool and carbon pool change estimates, bio-energy estimates, as well as future cutting possibilities. These results are presented in other publications and reports. Furthermore, the results can be calculated for many kinds of sub-national units. During NFI9, the detailed forest resource estimates and alternative cutting scenarios were published by Forestry Centre regions in 14 publications (e.g. Tomppo et al. 1998a; Hirvelä et al. 1998). Some of the results have been and are calculated by request. However, a sample-based inventory, where a certain minimum number of the observations are needed to keep the sampling errors at an acceptable level, also involves some limitations.

The need to intensify forestry created a demand for forest resource information for areas smaller than was possible with the typical NFI sampling intensities. Meanwhile, in the late 1980s, the appearance of new information sources, such as satellite images, stimulated further development the NFI. A new method called multi-source national forest inventory (MS-NFI) that utilises satellite images and digital map data, e.g. land-use data, elevation data and soil data, in addition to field measurements, has been operative since 1990. The aim of this multi-source inventory is to produce geographically localized information for small areas, e.g. for individual municipalities, which in Southern Finland typically have an area order of 10,000 ha. The MS-NFI9 method is described and municipality-level estimates are presented in Tomppo et al. (2008). These results and estimates in map form, both available in digital form, are employed by forestry authorities to assess municipality-level forest resources and cutting possibilities. Furthermore, forest industries use them to assess cutting possibilities by forest holdings. They are also used by nature conservation agencies to assess landscape-level biodiversity and the need and possibilities for forest protection. Researchers also use the results for diverse research purposes. The use of remote sensing has brought additional information sources but has not removed the need for precise ground measurements.

We hope that this book will provide tools, ideas and examples for decision-makers and policy planners as well as act as guide for the implementation of national forest or any large-area inventories.

References

Appelroth E, Heikinheimo O, Kalela EK, Laitakari E, Lindfors J, Sarvas R (1948) Julkilausuma. Metsätaloudellinen Aikakauslehti 11 (In Finnish)
Blomqvist AG (1881) Statistiska studier för utrörande af Finlands internationela ställning såsom skogproducerande stat. Finska Forstföreningens Meddelanden 2:91–124
Cajanus W (1913) Silmämääräinen metsän kuutiomäärän arvioimistapa (A method for ocular estimation of the growing stock of forests). Tapio 6:77–79 (In Finnish)
Ervasti S, Heikinheimo L, Holopainen V, Kuusela K, Sirén G (1965) The development of Finland's forests in 1964–2000. Silva Fennica 117(2):1–35
Ervasti S, Heikinheimo L, Kuusela K, Mäkinen VO (1970) Forestry and forest industry production alternatives in Finland, 1970–2015. Folia Forestalia 88:1–65
Gregoire TG, Valentine HT (2008) Sampling strategies for natural resources and the environment. Chapman & Hall/CRC, Boca Raton, London, New York
Gyldén CW (1853) Handledning för skogshushållare i Finland. Helsingfors
Hannikainen PW (1896) Suomen metsät kansallis-omaisuutemme (Finland's forests, our national treasure). The Finnish National Forest Service, Helsinki
Heikkinen J (2006) Assessment of uncertainty in spatially systematic sampling. In: Kangas A, Maltamo M (eds.) Forest inventory – methodology and applications, managing forest ecosystems, vol 10. Springer, Dordrecht, pp 155–176
Heikurainen L, Kuusela K, Linnamies O, Nyyssönen A (1960) Metsiemme hakkuumahdollisuudet. Pitkän ajan tarkastelua. Silva Fennica 110:1–52 (In Finnish)
Henttonen HM, Mäkinen H, Nöjd P (2009) Seasonal dynamics of the radial increment of Scots pine and Norway spruce in the southern and middle boreal zones in Finland. Can J For Res 39:606–618
Hirvelä H, Nuutinen T, Salminen O (1998) Valtakunnan metsien 9. inventointiin perustuvat hakkuumahdollisuusarviot vuosille 1997-2026 Etelä-Pohjanmaan metsäkeskuksen alueella. Julkaisussa: Etelä-Pohjanmaa. Metsävarat 1968-97, hakkuumahdollisuudet 1997-2026. Metsätieteen aikakauskirja – Folia Forestalia 2B:279–291 (In Finnish)
Ilvessalo Y (1923) Tutkimuksia yksityismetsien tilasta Hämeen läänin keskiosissa. Sahalahden ja Kuhmalahden pitäjien metsät. Untersuchungen über den Zustand der Privatwälder in den mittleren Teilen des Läns Tavastehus. Die Wälder der Gemeinden Sahalahti und Kuhmalahti. Acta Forestalia Fennica 26:1–137 (In Finnish with German summary)
Ilvessalo Y (1924) The forests of Finland. The forest resources and the condition of the forests. Communicationes ex Instituto Quaestionum Forestalium Finlandiae editae 9
Ilvessalo Y (1927) The forests of Suomi Finland. Results of the general survey of the forests of the country carried out during the years 1921–1924. Communicationes ex Instituto Quaestionum Forestalium Finlandie 11 (In Finnish with English summary)
Ilvessalo Y (1942) Suomen metsävarat ja metsien tila: II valtakunnan metsien arviointi (The forest resources and the condition of the forests of Finland: the Second National Forest Survey). Communicationes Instituti Forestalis Fenniae 30:1–446
Ilvessalo Y (1956) Suomen metsät vuosista 1921–24 vuosiin 1951–53. Kolmen valtakunnan metsien inventointiin perustuva tutkimus (The forests of Finland from 1921–24 to 1951–53.

A survey based on three national forest inventories). Communicationes Instituti Forestalis Fenniae 47(1):1–227

Ilvessalo Y (1961) Metsiemme kasvun toivetavoite. Metsälehti 48–49 (In Finnish)

Ilvessalo (1962) IV Valtakunnan metsien inventointi. 1. Maan eteläpuoliskon vesistöalueryhmät (Fourth national forest inventory. 1. Southern water system areas). MTJ 56.1. Valtioneuvoston kirjapaino, Helsinki, 112 p, 2 kpl

Ilvessalo (1963) IV Valtakunnan metsien inventointi. 2. Maan eteläpuoliskon metsänhoitolautakuntien alueryhmät (Fourth national forest inventory. 2. Southern forestry board districts) Summary in English. MTJ 57.4. Valtioneuvoston kirjapaino, Helsinki, 100 p

IPCC (2003) Good practice guidance for land use, land use change and forestry. In: Penman J, Gytarsky M, Hiraishi T, Krug T, Kruger D, Pipatti R, Buendia L, Miwa K, Ngara T, Tanabe K, Wagner F (eds.). Hayama: Intergovernmental Panel on Climate Change (IPCC). http://www.ipccnggip.iges.or.jp/public/gp/gpgaum.htm

Järvinen O, Kuusela K, Väisänen RA (1977) Effects of modern forestry on the number of breeding birds in Finland 1945-1975. Silva Fennica 11:284–294

Kangas A, Maltamo M (eds.) (2006) Forest inventory. Methodology and applications. Managing forest ecosystems, vol 10. Springer, Dordrecht

Kangas K, Luoto M, Ihantola A, Tomppo E, Siikamäki P (2010) Recreation-induced changes in boreal bird communities in protected areas. Ecol Appl 20(6):1775–1786

Kujala V (1964) Metsä- ja suokasvilajien levinneisyys- ja yleisyyssuhteista Suomessa. Vuosina 1951–1953 suoritetun valtakunnan metsien III linja-arvioinnin tuloksia. Über die frequenzverhältnisse der Wald- und Moorpflanzen in Finnland. Ergebnisse der III. Reichswaldabschätzung 1951–1953, 137 p (33 appendix maps)

Kuusela K, Salminen S (1969) The 5th national forest inventory in Finland. General design, instructions for field work and data processing. Commununicationes Instituti Forestalis Fenniae 69(4):1–72

Labau VJ, Bones JT, Kingsley NP, Lund HG, Smith WB (2007) A history of the forest survey in the United States: 1830-2004. US Department of Agriculture, Washington, DC, pp 1–82, FS-877

Langsæter A (1926) Om beregning av middelfeilen ved regelmessige linjetaksering. Meddelelser fra det norske Skogforsøksvesen 2(7):5–47

Langsæter A (1932) Nøiaktigheten ved linjetaksering av skog, I. Meddelelser fra det norske Skogforsøksvesen 4:431–563

Lehtomäki J, Tomppo E, Kuokkanen P, Hanski I, Moilanen A (2009) Applying spatial conservation prioritization software and high-resolution GIS data to a national-scale study in forest conservation. For Ecol Manage 258:2439–2449

Lindeberg JW (1924) Über die Berechnung des Mittelfehlers des Resultates einer Linientaxierung. Acta Forestalis Fennica 25

Lindeberg JW (1926) Zür Theorie der Linientaxierung. Acta Forestalis Fennica 31(6):1–9

Loetsch F, Haller KE (1964) Forest inventory, vol I. BLV Verlagsgesellschaft mbH, München

Loetsch F, Zöhrer F, Haller KE (1973) Forest inventory, vol II. BLV Verlagsgesellschaft mbH, München

Mandallaz D (2008) Sampling techniques for forest inventories. Chapman & Hall/CRC, Boca Raton, 256 p

Mattila E (1985) The combined use of systematic field and photo samples in a large-scale forest inventory in North Finland. Seloste: Systemaattisen ilmakuva- ja maastonäytteen yhteiskäyttö laajan metsäalueen inventoinnissa Pohjois-Suomessa. Communicationes Instituti Forestalis Fenniae 131:1–97

Mattila E (1996) Porojen talvilaitumet Suomen poronhoitoalueen etelä- ja keskiosissa 1990-luvun alussa. Metsätieteen aikakauskirja - Folia Forestalia 1996(4):337–357 (In Finnish)

MELA ja metsälaskelmat (2010) Timber production potentials and MELA-system. http://mela2.metla.fi/mela/. Accessed 7 Dec 2010 (In Finnish)

MERA III (1969) MERA III rahoitusohjelma. MERA III Forest Finance Committee, Helsinki

References

Metsä 2000 ohjelmajan tarkistustoiminkunta (1992) Metsä 2000 ohjelman tarkistustoimikunnan mietintö. Komiteamietintö 1992:5. Valtion painatuskeskus. Maa- ja metsätalousministeriö, Helsinki, 112 p (In Finnish). ISBN 951-47-3837-3

Metsä 2000 ohjelmajaosto (1985) Metsä 2000 – ohjelman pääraportti. Talousneuvosto, Helsinki (In Finnish). ISBN 951-46-8662-4

Ministerial Conferences on the Protection of Forests in Europe (MCPFE) (2007) State of Europe's Forests 2007: The MCPFE report on sustainable forest management in Europe. http://www.foresteurope.org/?module=Files;action=File.getFile;ID=485. Accessed 1 Nov 2010

Myllyntaus T, Mattila T (2002) Decline or increase? The standing timber stock in Finland. Environ Econ 41(2):271–288

National Board of Customs (2010) Statistics and information service. http://www.tulli.fi/en/finnish_customs/statistics/statistics_service/index.jsp. Accessed 7 Dec 2010

National Forestry Program (2010) The ministry of agriculture and forestry. http://www.mmm.fi/kmo/english/. Accessed 27 June 2011

Nyyssönen A (1998) Metsiemme kasvu: HKLN-ohjelman tavoitteet toteutumassa. Metsätieteen aikakausikirja 1:104–107

Östlind J (1925) Über die Berechnung des Mittelfehlers des Resultates einer Linientaxierung (rec.) Skogsvårdföreningens tidskrift. Stockholm

Östlind J (1932) Erforderlig taxeringsprocent vid linjetaxering av skog. Svenska Skogsföreningens Tidskrift 30:417–514

Päätös lääninmetsälautakuntien toiminnan lopettamisesta ja metsänhoitolautakuntien toiminnan aloittamisesta (1928) Decree 267/1928

Pakkala T, Hanski I, Tomppo E (2002) Spatial ecology of the three-toed woodpecker in managed forest landscapes. In: Korpilahti E, Kuuluvainen T (eds.) Disturbance dynamics in boreal forests: defining the ecological basis of restoration and management of biodiversity. Silva Fennica 36(1):279–288

Poso S, Kujala M (1971) Ryhmitetty ilmakuva- ja maasto-otanta Inarin. Utsjoen ja Enontekiön metsien inventoinnissa. Summary: A two-phase forest inventory method based on photo and field sampling as applied in Northernmost Lapland. Folia Forestalia 132:1–40

Poso S, Kujala M (1977) Menetelmä valtakunnan metsien inventointiin Pohjois-Suomessa. 54 s. Summary: A method for national forest inventory in northern Finland. Communicationes Instituti Forestalia Fenniae 93.1. Helsinki, Finland, 54 p

Räsänen PK, Pohtila E, Laitinen E, Peltonen A, Rautiainen O (1985) Metsien uudistaminen kuuden eteläisimmän piirimetsälautakunnan alueella. Vuosien 1978–1979 inventointitulokset (Summary: Forest regeneration in the six southernmost forestry board districts of Finland. Results from the inventories 1978–1979). Folia Forestalia 637:1–30

Raunio A, Schulman A, Kontula T (eds.) (2008) Suomen luontotyyppien uhanalaisuus. Suomen ympäristökeskus, Helsinki. Suomen ympäristö 8/2008. Osat 1 ja 2. Summary: The assessment of threatened habitat types in Finland. Finnish Environment Centre, Helsinki, Finland. Parts 1 and 2, 264+572 p

Reinikainen A, Mäkipää R, Vanha-Majamaa I, Hotanen JP (eds.) (2000) Kasvit muuttuvassa metsäluonnossa. Summary: changes in the frequency and abundance of forest and mire plants in Finland since 1950. Tammi, Helsinki, 384 p

Sarvas R (1944) Tukkipuun harsintojen vaikutus Etelä-Suomen yksityismetsiin. Referat: Einwirkung der Sägestämmeplenterungen auf di Priwatwälder Südfinnlands. Communicationes Instituti Forestalia Fenniae 33(1):1–268

Sarvas R (1950) Tutkimuksia Perä-Pohjolan harsimalla hakattujen yksityismetsien uudistamisesta (Summary: Investigations into the natural regeneration of selectively cut private forests in northern Finland). Communicationes Instituti Forestalia Fenniae 38(1):1–95

Schreuder HT, Gregoire TG, Wood GB (1993) Sampling methods for multiresource forest inventory. Wiley, New York, 446 p

Statistics Finland (2010) Greenhouse gas emissions in Finland 1990-2008. National inventory report to under the UNFCCC and the Kyoto protocol. http://www.stat.fi/tup/khkinv/fin_nir_20100525.pdf. Accessed 1 Nov 2010

Tomppo E, Henttonen H, Korhonen KT, Aarnio A, Ahola A, Heikkinen J, Ihalainen A, Mikkelä H, Tonteri T, Tuomainen T (1998a) Etelä-Pohjanmaan metsäkeskuksen alueen metsävarat ja niiden kehitys 1968-97. Julkaisussa: Etelä-Pohjanmaa. Metsävarat 1968-97, hakkuumahdollisuudet 1997-2026. Metsätieteen aikakauskirja – Folia Forestalia 2B:293–374 (In Finnish)

Tomppo E, Katila M, Moilanen J, Mäkelä H, Peräsaari J (1998b) Kunnittaiset metsävaratiedot 1990-94. Metsätieteen aikakauskirja – Folia Forestalia 4B:619–839 (In Finnish)

Tomppo E, Henttonen H, Tuomainen T (2001) Valtakunnan metsien 8. inventoinnin menetelmä ja tulokset metsäkeskuksittain Pohjois-Suomessa 1992–94 sekä tulokset Etelä-Suomessa 1986–92 ja koko maassa 1986–94. Metsätieteen aikakauskirja 1B:99–248 (In Finnish)

Tomppo E, Haakana M, Katila M, Peräsaari J (2008) Multi-source national forest inventory – methods and applications, managing forest ecosystems, vol 18. Springer, Dordrecht, 374 p. ISBN 978-1-4020-8712-7

Tomppo E, Schadauer K, McRoberts RE, Gschwantner T, Gabler K, Ståhl G (2010a) Introduction. In: Tomppo E, Gschwantner T, Lawrence M, McRoberts RE (eds.) National forest inventories – pathways for common reporting. Springer, Dordrecht, pp 1–18. ISBN 978-90-481-3232-4

Tomppo E, Gabler K, Schadauer K, Gschwantner T, Lanz A, Ståhl G, McRoberts R, Chirici G, Cienciala E, Winter S (2010b) Summary of accomplishments. In: Tomppo E, Gschwantner T, Lawrence M, McRoberts RE (eds.) National forest inventories: pathways for common reporting. Springer, Dordrecht, 672 p

Tuhkanen S (1980) Climatic parameters and indices in plant geography, vol 67, Acta Phytogeographica Suecica. Almqvist & Wiksell International, Stockholm, 105 p

von Berg E (1858) Kertomus Suomenmaan metsistä 1858. In: Leikola M (ed.) Edmund von Berg: Kertomus Suomenmaan metsistä 1858, sekä Kuvia suuresta muutoksesta. Kustannusosakeyhtiö Metsälehti, Gummerus Kirjapaino Oy, Jyväskylä, 1995, 3 p (In Finnish)

Chapter 2
Design and Measurements

National Forest Inventory of Finland (NFI) has evolved from a line survey to a plot-based inventory (Sect. 1.3). Simulated sampling was introduced as a tool for planning the design during NFI8 and it was fully adopted in NFI9 (Sect. 2.1). The main assessment units in the field work are stands and individual trees, although some of the variables are assessed on circular plots with fixed radius (Sect. 2.2). Section 2.3 explains how the centre points of the sample plots were located in the field.

A fairly detailed and comprehensive description of the variables assessed in NFI9 is given in Sects. 2.4–2.20. For some variables, this is essential for the interpretation of the results reported in Chap. 4 and the Appendix tables, but the description also contains many variables for which results are not presented. The aims are to demonstrate the full extent and complexity of a real-life NFI, to give a realistic picture of the amount of information achieved with the reported effort (Sect. 2.24), and to provide the kind of details that might be useful to those actually implementing an inventory. This text, in particular Tables 2.1–2.34, can also serve as a reference to the users of NFI9 data, which is facilitated by listing the codes actually used in the NFI database.

Section 2.21 presents the measurement devices employed and Sect. 2.22 details a correction to height and height increment measurements which was found necessary after the first year's experiences with a new instrument. The training of the field crews and the control measurements are described in Sect. 2.23.

2.1 Field Sampling Design

The general approach in recent NFIs has been to group sample plots into clusters, the measurement of which should amount to approximately one day's work. The plots of one cluster should be so close to each other that it is feasible to move between them on foot, but sufficiently far apart to avoid too many cases, where one stand contains multiple plots. With the exception of the sparsely forested northernmost part of the country (see Sect. 2.1.4), the clusters have been spread out systematically, in the form of a rectangular grid. The design has been adapted to the large-scale spatial variability in the forests so that the sampling is sparser in the north, where the stands are larger and the variation in the stock is smaller. Also the administrative regions, for which inventory results are required, are larger in the north.

Forestry centres (Fig. 2.1a) are the most important administrative units, for which NFI results are reported. The measurements of NFI9 also proceeded by forestry centres.

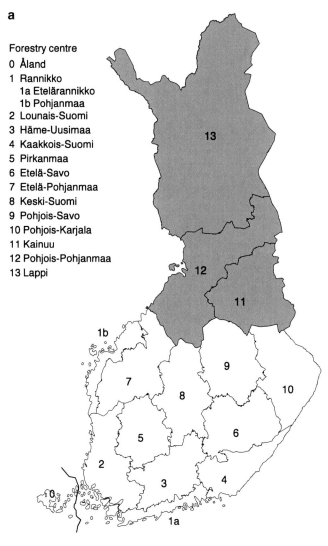

Fig. 2.1 (**a**) Forestry centres 1.1.2003 (North Finland grey). Digital map data: © National Land Survey of Finland, licence No. 6/MYY/11. (**b**) Sampling density regions and the locations of the field plot clusters in the 9th National Forest Inventory. In the entire country, there were 81,249 sample plot centres on land, 67,264 on forestry land, 62,266 on forest land and poorly productive forest land, and 57,457 on forest land

2.1 Field Sampling Design

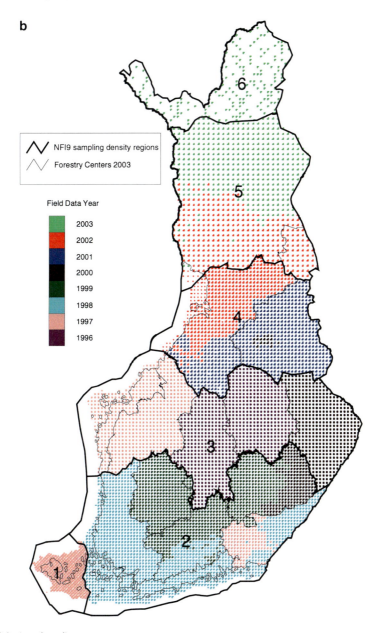

Fig. 2.1 (continued)

The adopted inventory design (e.g., the density of clusters, and the number of and distances between the plots in a cluster) varied between *sampling density regions* (Fig. 2.1b). The borders between these regions generally followed those of the forestry centres. The exceptions were caused by changes in the administrative units.

Permanent plots were introduced in the later stages of NFI8, when North Finland was measured. Three permanent plots were established in each cluster and re-measured in NFI9. However, a new strategy was adopted in NFI9, where every fourth cluster was made permanent except in Åland (Fig. 2.2). Because the workload per plot is heavier on permanent than temporary clusters, a smaller number of plots was assigned for a permanent cluster than for a temporary one.

2.1.1 Sampling Simulation

The sampling designs in each region were chosen on the basis of sampling simulation studies (Henttonen 1991, 1996; Tomppo et al. 1998a, 2001) using digital maps of land use and of the volume of growing stock. Pixels in the digital maps were repeatedly sampled (1,000–2,000 repetitions) according to each potential design, and the *simulated error*, the standard deviation of the sample means, was assumed to anticipate the sampling error associated with that design in relation to the competitive designs. The implementation varied by region depending on the available data (for details, see Sects. 2.1.2–2.1.4), but the general idea was to choose the design with the smallest simulated error among those satisfying the budget constraints. The time consumption of the different phases, such as walking to a cluster, moving from one plot to the next, measuring a tree, conducting the stand level assessment, and mapping the trees on the permanent plots, were partly taken from a study by Päivinen and Yli-Kojola (1983) and partly collected from field crews and earlier experiences (Henttonen 1991; Tomppo et al. 2001).

2.1.2 South Finland

Test areas of 60 km × 60 km were selected from sampling density regions 2 (13 areas) and 3 (8 areas). The sampling was simulated on a thematic map of land use and the volume of growing stock on forestry land produced by the multi-source NFI using digital map data, satellite images and NFI8 field data (Tomppo et al. 1998b, 2008). In addition to the aim of measuring South Finland in three field seasons, an upper limit of approximately 3% was set to the simulated relative sampling error of the mean volume estimates for the test areas, based on the simulated error obtained with the design of NFI8.

It was found that the 'optimum' design depended on, for instance, the distribution of forest land and the heterogeneity of the forests and therefore varied from south to north (Fig. 2.3). An L-shaped cluster with 14 plots was selected for region 2 (10 plots on the permanent clusters) and a rectangle with 18 plots for region 3 (14 plots on a permanent cluster). The distances between the clusters were 6 km × 6 km in region 2 and 7 km × 7 km in region 3 (Fig. 2.2b, c). For sampling density region 1 (the county of Åland) with land area much smaller than that of the

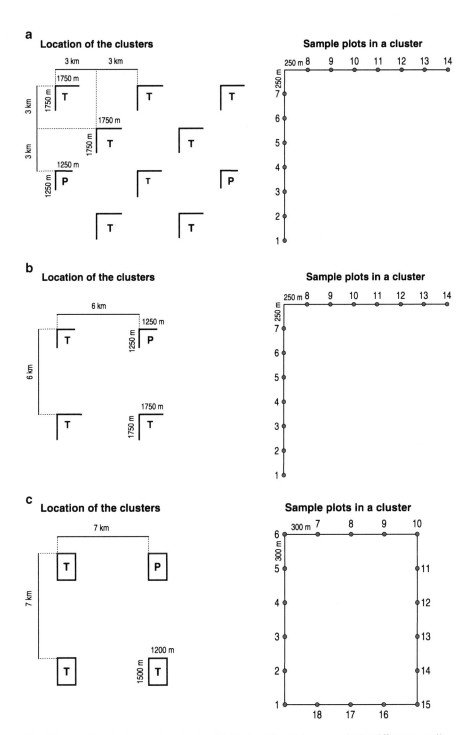

Fig. 2.2 Sampling design employed in the 9th National Forest Inventory in the different sampling density regions of Fig. 2.1b: (**a**) region 1; (**b**) region 2; (**c**) region 3; (**d**) region 4 (in region 5, the design is same but the distance between clusters is 10 km × 10 km); (**e**) region 6, stratified sampling was applied in region 6

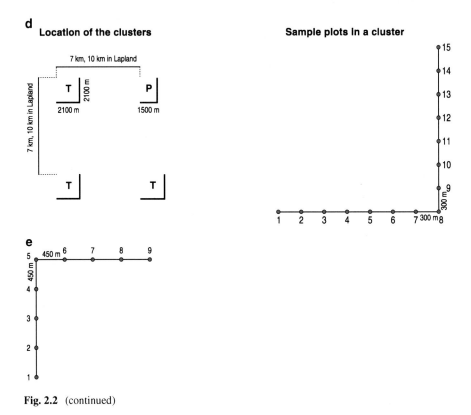

Fig. 2.2 (continued)

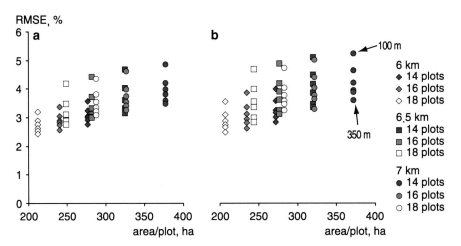

Fig. 2.3 Estimated root mean square errors (%) of total volume for alternative sampling designs in an area of 60 km × 60 km in South Finland. Cluster shape (**a**) is a semi-rectangle and shape (**b**) is rectangle

other forestry centres, the sampling density was approximately doubled by adding an extra row of clusters between each row in the design of region 2 (Fig. 2.2a).

2.1.3 North Finland (Excluding North Lapland)

The design of NFI9 in sampling density regions 4 and 5 was dictated by that of NFI8, owing to the need to re-measure the permanent plots. Thematic maps of the multi-source inventory were not available when the design of NFI8 in region 4 was determined. Instead, sampling was simulated on the maps of the land use and growing stock classification by the National Land Survey, which was based on Landsat 5 TM data and diverse field data sources. For region 5, NFI7 data was used to develop a regression model for predicting the mean volume for each pixel of Landsat 5 TM image.

The aim of the design was to measure each of the two regions (with a land area of approximately 5.7 million ha in region 4 and 6.5 million ha in region 5) in one field season. Among those designs fulfilling this criterion, two designs yielded the smallest simulated sampling errors: a cluster with 15 plots in L-form with plot distance of 300 m, and a rectangular-shaped cluster with 13 plots and plot distances of 400 m. The distance between clusters was 7 km in region 4 and 10 km in region 5. The cluster with 15 plots was selected for both regions in order to allocate more time for field measurements.

The designs were kept the same for NFI9 as in NFI8, except that the direction of the sides of the L-form was changed in such a way that the three permanent plots could be re-measured and the number of the plots in permanent clusters was reduced to 11 (Fig. 2.2d).

2.1.4 North Lapland

Double sampling for stratification (Cochran 1977, sec. 12.2) was employed in North Lapland (sampling density region 6). The proportion of forest land is low, there are large areas without any forest land, the occurrence of forests is patchy, and the road network is sparse. These facts support the use of methods other than systematic sampling when the efficiency requirements are taken into account. A special study was conducted to determine an efficient statistical sampling design (Henttonen 2003).

The study began with measurements of 50 m × 50 m test areas to compare the efficiencies of the alternative field plot sizes and shapes, as well as possible field plot cluster sizes, and to determine the 'optimal' field plot and field plot cluster. In total, 18 test areas were measured, together with the planar coordinates of the trees ($d_{1.3}$ at least 2.5 cm), as well as breast height diameter, height and upper diameter at a height of 6 m (for trees taller than 8 m). The plot alternatives tested were angle count plots with several basal area factors (see Sect. 2.2.1 for details on angle count plots) and concentric fixed radius plots with different lengths of the radius. Furthermore, a single plot versus a plot split into two or three sub-plots was studied. Judgements between options were made on the basis of time consumption and the differences between the 'true' mean values over all test areas and their estimates obtained by simulating one plot to each test area. The characteristics employed were volume per hectare and the number of trees per hectare. The number of boundary trees was also taken into consideration when estimating time consumption, in addition to the other phases in measuring a field plot and a field plot cluster.

As a result, an angle count plot with a basal area factor of 1.5 was selected. The selected cluster, using a constraint of one working day for its measurements, consisted of nine plots with a distance of 450 m of two adjacent plots (Fig. 2.2e).

In North Lapland, the measurement costs are dominated by the time required to reach a plot from the closest road. The first phase sample consisted of a regular (systematic) grid of the clusters with a spacing of 7 km × 7 km. The proportion of unproductive land and the measurement costs were the main criteria when selecting the second phase sample. The clusters were stratified for the second phase sample as follows (note that all the plots in a cluster belonged to a same stratum): for each cluster, the proportion of the pixels on unproductive land was calculated using a window of 9×9 pixels around each plot. The predictions of land classes based on the multi-source national forest inventory principle were employed (Tomppo et al. 1998b). To reduce the effect of the estimation error, the mean volume of the growing stock was also calculated for the same pixels using the same data source. Further, the 95% confidence interval of the prediction of the effective long term temperature sum was used for each plot. The observations of the Finnish Meteorological Institute and kriging interpolations were used (Henttonen 2003). Further, topographic maps from the National Land Survey were used for assessing the costs to reach a cluster.

Denoting the prediction of the effective temperature sum by d.d. and its 95% upper confidence limit by $u_{t,95}$ and the proportion of the pixels of a cluster on unproductive land (as given presented above) by p_u, the six strata were:

1. $0\% \leq p_u \leq 25\%$, difficult access to the cluster, and $u_{t,95} > 550$ d.d. for at least one plot of the cluster.
2. $0\% \leq p_u \leq 25\%$, easy access (distance to road not more than 1 km without major water bodies between) to the cluster, and $u_{t,95} > 550$ d.d. for at least one plot of the cluster.
3. $25\% < p_u \leq 50\%$, and for at least one plot of the cluster $u_{t,95} > 550$ d.d.
4. $50\% < p_u \leq 90\%$, the mean volume >14 m³/ha for at least one pixel of the cluster, and $u_{t,95} > 550$ d.d. for at least one plot of the cluster.
5. $50\% < p_u \leq 90\%$, and the mean volume ≤14 m³/ha for all pixels of the cluster, plus the clusters with $90\% < p_u \leq 100\%$, if the cluster does not belong to stratum 6, plus the clusters with $u_{t,95} \leq 550$ d.d. for all plots, if the cluster does not belong to stratum 6
6. The clusters with $u_{t,95} \leq 400$ d.d. for all plots,
   plus the clusters with $u_{t,95} \leq 550$ d.d. for all plots and the mean volume ≤10 m³/ha for all pixels,
   plus the clusters with $90\% < p_u \leq 100\%$ and the mean volume ≤8 m³/ha for all pixels.

2.2 Assessment Units

The clusters in strata 1–4 were at least partly on combined forest land and poorly productive forest land. The plots of strata 5 and 6 are probably either on unproductive land or on poorly productive land with a low volume of growing stock. The 95% upper confidence limit of the effective temperature sum was $u_{t,95} \leq 550$ d.d. for all the plots of these two strata.

The number of clusters for the second phase sample was selected on the basis of the available budget and was set to 180 clusters. These clusters were allocated to the six strata using an optimal allocation (e.g. Cochran 1977). The between-cluster standard deviations of the mean volume within each stratum were used, as well as the assessed relative measurement costs of a cluster. The relative cost was 0.8 for the clusters with easy access and 1 for those with difficult access.

2.2 Assessment Units

2.2.1 Angle Count Plots

Tally trees were selected using angle count sampling (see Sect. 2.16 for the definition of a tree and the tally tree measurements). A tree with a breast height (1.3 m) diameter of $d_{1,3}$ is included in an *angle count plot* (Bitterlich plot), if its distance from the *sample plot centre* (a point determined by the sampling design, Sect. 2.1) is at most $r = 50 d_{1,3} / \sqrt{q}$, where q is the *basal area factor* (Fig. 2.4). In NFI9, q was 2 in South Finland and 1.5 in North Finland, where trees are smaller and sparser. However, the tally tree plots were restricted to maximum radii of 12.52 m in South Finland and 12.45 m in North Finland corresponding to breast height diameters of

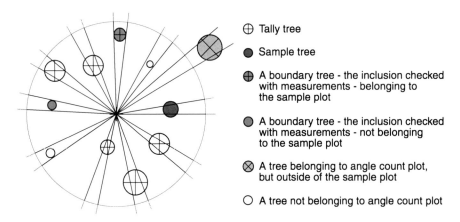

Fig. 2.4 A sample plot as used in the 9th National Forest Inventory. The maximum radius for trees to be counted was 12.52 m in South Finland ($q=2$) and 12.45 m in North Finland ($q=1.5$). Every seventh tree was measured as a sample tree, except in North Lapland where every fifth was measured

35.4 and 30.5 cm, respectively. Reducing the radius of a sample plot detracts very little from the precision of the estimates, but it does ease the amount of fieldwork noticeably in some cases, as the number of divided sample plots, i.e. sample plots belonging to two or more stands, decreases. The use of maximum distance also reduces systematic errors caused by unobserved trees, located a long distance from the plot centre and behind other trees. Where a relascope could not be used reliably, inclusion was checked by measuring the distance and diameter of the tree at a height of 1.3 m.

A subsample of tally trees was selected as *sample trees*, from which many additional characteristics were assessed (see Sect. 2.16, Table 2.27). Every 7th tally tree was measured as a sample tree, except in North Lapland where it was every 5th tally tree.

2.2.2 Stands

Most of the variables described in Sects. 2.4–2.14 were assessed at the level of a stand. For the Finnish NFI, a *stand* is defined as a connected land and its possible growing stock that is judged by the field crew leader to be homogeneous with respect to certain variables. Main criteria for delineating stands are administrative boundaries, land use, principal site class, site fertility, and the structure of the growing stock, such as maturity class and tree species composition, as well as the extent of accomplished or needed silviculture and cutting regimes. All of these variables are discussed in detail in the following sections. Note that, for simplicity, the term 'stand' may also refer to a patch of non-forestry land, such as an agricultural field or road, or even to a water body. This is in contrast to the use of the word in some other literature, where stand refers to the growing stock only.

The NFI sample included all *centre point stands*, i.e., stands containing at least one of the sample plot centres, and those *bi-stands*, i.e., stands not containing a plot centre, that contained tally trees.

2.2.3 Other Assessment Units

In case of a multi-storied tree stand, some characteristics concerning the growing stock were assessed by *tree storey* (Sect. 2.11). Furthermore, some variables were assessed within a *fixed radius circular sample plot* around the plot centre; namely, topography and coverage of peatland mosses, as described in Sect. 2.6, soil variables of Sect. 2.7, and key habitats of Sect. 2.15. To enable unbiased estimation of the area of key habitats (Sect. 3.1), the area of combined forest land, poorly productive forest land and unproductive land inside the 30 m circle, the key habitat plot, was assessed (as tenth parts).

Dead wood was measured on centre point stands on forest land and poorly productive forest land within a circle with a radius of 7 m, with the exception of the first year of NFI9, 1996, when forestry centres 8 (Keski-Suomi) and 9 (Pohjois-Savo) were measured and the radius was 12.52, see Sect. 2.20. For unbiased estimation of the amount of dead wood, it was necessary to assess the area of the centre point stand inside the dead wood plot (see Sect. 3.4). Fixed-radius plots were also used to sample large trees of keystone species (Sect. 2.18) and to assess the number of tree species (Sect. 2.19).

2.3 Locating the Field Plots

The first attempt to locate the field plots with GPS (Global Positioning System) devices was carried out in the first year of NFI9 (1996). A Garmin GPS 45 single channel device was used. The RDS signal provided by the Finnish Broadcasting company was used to correct the obtained locations and a minimum of three satellite observations was required from the GPS measurements stored in the field. The obtained locations were not precise enough and the collection of GPS data fulfilling the above requirements tended to take too much time. In a study by Rouvinen et al. (1999), the computed GPS locations from the above GPS data were compared to the field plot locations using map measurements, precision compass and measuring tape, as described below, and the minimum average difference obtained between GPS and map based measurements was 24 m in a selected set of field plots.

Thus the field plots in South Finland (1996–2000) were located using *map measurements*. First, a fixed point close to the line between two plots was selected. Estate corner marks and roads crossings or ditches were often used as fixed points. The distance from the fixed point to the line between two plots and distance from this point to the nearest plot was measured from the map and with a measuring tape and a precision compass in the field. Once the location of one plot had been defined, the field team moved from one plot to the next using a compass and measuring tape. The map had to be checked frequently to make sure that orientation was correct. If the difference between the real location and the expected location was more than 30 m, a new fixed point had to be selected.

By 2001, when NFI9 proceeded to North Finland, a sufficiently precise GPS receiver (Trimble ACE II) had been purchased. The GPS system was assembled at Metla installing the receiver into a backpack together with battery and cabling. A field computer was used for display and storage of GPS data. The external antenna was mounted on the backpack.

In North Finland, GPS was used to locate sample plots on forestry land and in other land use classes when the distance from the field plot to the nearest forestry land polygon was less than 20 m. GPS was used both for navigating to the predefined plot location, and for recording the precise location of the plot. Emphasis was placed on the latter aspect because it is very important in multi-source inventory to know the exact locations of the plots.

The field computer software computed the location of the next field plot and guided the crew to the plot. At the plot, GPS data (including the raw data from the receiver) was collected for 2 min to compute a more accurate plot location.

To improve accuracy, the collected GPS data was corrected using reference data from two sources. For the years 2000–2002 the reference data was obtained from the Geodetic Institute of Finland (station in Joensuu for 2000 and station in Oulu for 2001–2002). In 2003, publicly available data from the EUREF Permanent Network Sodankylä station was used. The precise GPS satellite orbits from International GNSS Service (IGS) were used in 2003. The post-correction software was created at Metla.

It is very difficult to assess the accuracy of a GPS system in realistic forest conditions. It is well known that the expected errors are much larger than the errors given by manufacturers for favourable conditions. No systematic accuracy studies have been made with the GPS system used in NFI9. Comparisons with other systems suggest that most of the computed locations are within 6 m of the correct location under "average forest conditions". However, errors larger than 10 m may occur in some cases. The accuracy of field plot locations based on map measurements is poorer. A rough estimate of the average location error when using maps was 20 m (Halme and Tomppo 2001). This is caused by map error and the errors in using the compass and tape.

2.4 Administrative Information

The administrative information includes the municipality, village, the register number of the holding, ownership information (Table 2.1), and possible restrictions on wood production (Tables 2.2 and 2.3). NFI covers the forest of all ownership groups, as well as wood production forests and protected forests of different protection categories. The forest holding register numbers were obtained before field work from the National Land Survey of Finland, and ownership information from databases of the Population Register Finland and the National Land Survey.

Table 2.1 Ownership categories

1	Private forest owners, farmers and other private owners separated from other data sources for special studies if needed
2	Forest companies
3	Other companies
4	Metsähallitus (main administrator of state owned forests)
5	Other areas owned by the state
6	Jointly owned forests
7	Municipalities, religious communities (parishes), and other communities
8	Other undivided ownership, e.g. estates of deceased

2.4 Administrative Information

Table 2.2 Main categories of restrictions on wood production

0	Areas with no restrictions
1	Areas where wood production was restricted by the Nature Conservation Act, e.g., strict nature reserves and national parks, protected herb-rich forest areas.
2	Areas where wood production was restricted by other laws, e.g., wilderness reserves, recreation areas owned by the state, archaeological sites.
3	Areas where wood production was restricted by authorities responsible for management, e.g., other than statutory protection areas, breeding and research forests, military areas.
4	Areas where wood production was restricted by government resolution including areas reserved for nature protection.
5	Areas where wood production was restricted by land-use planning, e.g., regional and local land use plans, local detailed plans.
6	Areas where wood production should be restricted, e.g. due to the special location, amenity values or habitat of threatened species.

Table 2.3 Categories of the intensity of restrictions

–	No restrictions on wood production.
1	No forest management activities permitted.
2	Areas where only activities to maintain or enhance nature values or biodiversity were permitted.
3	Areas where only specific forest management activities were permitted, e.g. single tree harvesting.
4	Areas under temporary preservation orders; e.g., areas allocated to conservation.
5	Areas requiring a permit to conduct activities, e.g. land-use planning areas.
6	Areas where wood production was affected by the closeness of other land use classes than forestry land (e.g. the closeness of arable land or power line). Fellings can be more intensive than usual.
7	Areas where forest land drainage was not permitted (felling was permitted).
8	Areas where the forest regime did not restrict wood production.

The wood production restrictions were obtained from various sources, e.g., from documents concerning the decisions to establish nature reserves, and maps of the land-use plans as well as the nature conservation databases of the Finnish Environment Institute and Metsähallitus. New restrictions identified during the field work were also recorded.

Category 6 in Table 2.2 was very diverse and it diverged from the other categories since the detection of these areas was based only on field observations by the field crew leader and not on documentation, as in the other cases. Within each category the intensity of the restriction varied according to the management regime. For example, in category 2, the sub-category "wilderness reserves" was divided into areas where all forestry activities were forbidden and areas where only specific forestry activities were permitted. Eight categories for permitted forest management activities were defined according to the regimes by sub-categories (Table 2.3).

2.5 Land-Use and Classification of Forestry Land

Both the national land-use classification (Table 2.4; Tomppo et al. 1998a) and the UNECE/FAO Temperate and Boreal Forest Resources Assessment classification, TBFRA 2000 (Table 2.5; UNECE/FAO 2000) were employed in NFI9. National land-use classes 1–4 make up what is called *forestry land*. The national definitions of land-use classes 1–4 have been employed since NFI5 (1964–1970, Kuusela and Salminen 1969).

Table 2.4 The national land-use classes in NFI9

1	*Forest land*: stocked or temporarily unstocked land with potential capacity to produce a mean annual increment of at least 1 m³/ha of stem wood over bark over the prescribed rotation under the most favourable stock conditions. Parks and yards are excluded.
2	*Poorly productive forest land*: stocked or temporarily unstocked land with potential capacity to produce a mean annual increment of 0.10–0.99 m³/ha of stem wood over bark. Parks and yards are excluded.
3	*Unproductive land*: either naturally treeless land or has the potential capacity to produce a mean annual increment of less than 0.10 m³/ha of stem wood over bark.
4	*Other forestry land*: forestry roads, forest depots and camp lots, small gravel and peat soil pits and game fields within forestry land.
5	*Agricultural land.*
6	*Built-up land*: urban areas, buildings etc.
7	*Land required by transport infrastructure.*
8	*Land under power lines.*
A	*Inland water bodies*: e.g., lakes and rivers with a width of at least 5 m.
B	*Salt water.*

Table 2.5 Land classes according to the FAO definitions (the definitions of a tree, shrub and bush in Sect. 2.16)

1	*Forest*: land with a tree crown cover (or equivalent stocking level) of more than 10% and an area of more than 0.5 ha. The trees should be able to reach a minimum height of 5 m at maturity in situ. Young natural stands and all plantations established for forestry purposes which have yet to reach a crown density of 10% or tree height of 5 m are included under forest, as are areas normally forming part of the forest area which are temporarily unstocked as a result of human intervention or natural causes but which are expected to revert to forest. For linear formations, a minimum width of 20 m is employed. Parks and yards, for example, are excluded regardless of whether they would meet the definition of forest land (UNECE/FAO 2000).
2	*Other wooded land*: land with a tree crown cover (or equivalent stocking level) of 5–10% of trees able to reach a height of 5 m at maturity in situ; or a crown cover (or equivalent stocking level) of more than 10% of trees not able to reach a height of 5 m in situ (e.g. dwarf or stunted trees) and shrub and bush cover. The area must be at least 0.5 ha in size and, in the case of linear formations, a minimum width of 20 m.
3	*Other land*: land not classified as forest land or other wooded land as they are defined above.

2.6 Site Variables

The *specification of land-use class* indicates, e.g., the location of the stand with respect to the surrounding land, or the area of the stand, if they were affecting the wood production. An example was a small forest stand surrounded by non-forest land. This information is useful when considering cutting possibilities.

The *change of the previous land-use class* and the *time since the change* describe any forestry land changes and changes from forestry land to other land-use classes and vice versa. In this way, changes which were small in area could be estimated more accurately with temporary plots than by using the differences of the area estimates. The gain is similar to that of using permanent plots. A possible failure to recognise changes several years back is a problem. The change within forestry land and changes from forestry land classes to other land-use classes were identified only during the last 10 years, and the changes from other land-use classes to forestry land classes during the last 30 years.

FAO TBFRA 2000 definitions (Table 2.5) were employed together with national definitions in field measurements starting in the summer of 1998 because of the need to employ FAO definitions in international reporting. The FAO definition of forest includes national forest land, a part of the poorly productive forest land, and forest roads of the national classification. The FAO classifications for the first 2 years of NFI9, 1996 and 1997, were derived from other stand level variables. In 1998–2002, the FAO land categories were assessed in the field only for the national land class of poorly productive forest land. It was assumed that all stands on national forest land were included in FAO forest while the stands belonging to national unproductive land belonged to neither FAO forest nor FAO other wooded land. In 2003, assessments based on FAO definitions were made for all centre point stands if the stand belonged to forest land, poorly productive forest land or unproductive land. Assessments were also made for those bi-stands on poorly productive forest land from which tally trees were measured. To support the assessments of the land categories based on FAO definitions, the actual canopy cover was assessed when the plot centre was on national forest land, poorly productive forest land or unproductive land. This assessment was adopted in 1998. The cover was assessed from a full circle with a radius of 12.52 m in South Finland and 12.45 m in North Finland, also when a circle possibly included other land classes than forest land and other wooded land.

2.6 Site Variables

For each sample plot centre, the *elevation* (dm above sea level) was obtained from a digital elevation model provided by the National Land Survey of Finland, and the average *effective temperature sum* (d.d.) over the 30-year period 1951–1980 was estimated as described in Ojansuu and Henttonen (1983).

The principal site class divides forest land, poorly productive forest land and unproductive land into mineral soil and peatland site classes (Table 2.6). A site is classified as *peatland* if the organic layer is peat or if more than 75% of the ground vegetation is peatland vegetation. Otherwise, the site is *mineral soil*.

Table 2.6 Principal site classes

1	Forest on mineral soil
2	Spruce mires
3	Pine mires
4	Treeless peatland (open bogs and fens)

Table 2.7 Site class specifications

0	Genuine peatland or mineral soil
1	Peatland with features of a forest on mineral soil
2	Spruce swamp features
3	Pine peatland features
4	Open bog or fen features
5	Eutrophic brown moss fen features
6	Naturally or artificially forested former non-forestry land

Table 2.8 Site fertility classes

1	Herb rich sites, eutrophic mires and fens and corresponding drained peatlands
2	Herb rich heath forests, mesotrophic mires and fens and corresponding drained peatland forests
3	Mesic forests on mineral soil and meso-oligotrophic natural and drained peatlands
4	Sub-xeric forests on mineral and oligotrophic natural and drained peatlands
5	Xeric forests on mineral soil and oligo-ombrotrophic natural and drained peatlands
6	Barren forests on mineral soil and *Sphagnum fuscum* dominated (ombrotrophic) natural and drained peatlands
7	Rocky and sandy soils and salt marsh (alluvial lands)
8	Summit and fell forests

Drained peatlands that have been open peatlands in their natural states but forested either naturally or by artificial regeneration are also classified as spruce- or pine-dominated peatlands.

Site class specification (Table 2.7) separates genuine and mixed classes. A site is classified as mixed if two different types are clearly discernible. On mineral soils, only codes 2, 3, and 6 are appropriate.

The site fertility classification (Table 2.8) in the Finnish NFI is based on the composition of ground vegetation on the site. The classification was created in the beginning of the twentieth century and further developed for forestry and NFI purposes (Cajander 1909, 1926; Kalela 1961, 1970). On the basis of the definition, a site category describes the vegetation composition when the growing stock is at the mature state and natural tree species composition is natural for the site. The large area climatic variation, as well as treatments such as accomplished cuttings, age of stand and variations in tree species composition, cause variations in the composition of the ground vegetation within the site category (e.g. Cajander 1949). Studies that analyse the effect of climatic variation on ground vegetation are summarised in Kalela (1961). Forest and peatland site types used in NFI refer to the forest vegetation zones (e.g. Lehto and Leikola 1987).

2.7 Soil Variables

Table 2.9 Site fertility class specification

0	No specification.
1	Patterned fen (>30% of the site covered with relict glacial flakes and pools).
2	*Molinia caerulea* as one of the dominant species.
3	*Sphagnum fuscum* dominated hummocks cover (>30% of the surface).
4	Inundation (that exceeds the normal spring flooding), and significant nutrient input by surface waters or by ground waters (the site eutrophy by surface waters is present (water-front, stream, spring etc.)). The vegetation includes flood meadow species such as *Alnus glutinosa, Salix* spp., *Calamagrostis canescens, C. purpurea*, tall sedges (*Carex* spp.), *Calla palustris, Caltha palustris, Phragmites australis* and *Filipendula ulmaria*. Even on drained sites, the surface water effect is visible in the vegetation by the presence of meadow species and the absence of forest species. Also sites that experience short-term eutrophying inundations are included in this class.
5	Shallow peat layer (<30 cm).
6	Mineral soil forest of the *Pyrola* type.

In national land-use classification (Table 2.4), all stands on mineral soil with site fertility class in 1–6 were classified as forest land. Class 7 can be forest land, poorly productive forest land, or unproductive land, and class 8 either poorly productive forest land or unproductive land. The site fertility class specification (Table 2.9) was specifically employed for mires, fens and bogs to specify the site type and its wood production potential.

Wood production potential is also described by topography and coverage of peatland mosses. *Topographical information* describes the variation of elevation and slope of the plot and its surroundings, mainly from the hydrological point of view. The unit of observation was a circle with a radius of 20 m around the sample plot centre. An assessment was made only if the plot centre was on forest land or poorly productive forest land.

Coverage of peatland mosses describes the moisture condition of the sample plot. This variable was adopted in NFI9 and was estimated on forest land, poorly productive forest land and unproductive land based on a circle with a radius equal to the maximum radius of the tally tree plot (12.52 m in South Finland and 12.45 m in North Finland). On mineral soils, mosses considered as peatland mosses were *Sphagnum* mosses and *Polytrichum commune*. On peatlands, open water surfaces (for example patterned fens and water depressions in spruce swamps) and other visibly wet surfaces with coverage of small true mosses (*Mniaceae, Calliergon* spp. and *Limprichtia* spp. etc.) were also included in the coverage estimate.

2.7 Soil Variables

Soil variables were assessed by stand on forest land, poorly productive forest land and unproductive land, but the assessment was restricted to a circle with a radius equal to the maximum radius of the tally tree plot. Measurements were taken at the points of a square grid with 4 m spacings (Fig. 2.5).

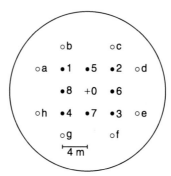

Fig. 2.5 Measurement points for soil variables; '0' is the sample plot centre, and primary measurement points were selected among points '1' to '8' (see descriptions of each variable for details). In a plot intersecting more than one stand, supplementary points (a–h) could be used to yield a sufficient number of points per stand

Table 2.10 Soil types

0	Organic soil
1	Bedrock
2	Stony soil, boulder field
3	Glacial till
4	Sorted soil

The soil variables employed in NFI9 were: soil type, mean grain size, depth of the soil, stoniness, decomposition stage of surface peat, organic layer type and the depth of the organic layer. Stoniness and decomposition stage of surface peat were assessed only on permanent plots in NFI9. The depth of the organic layer was measured down to 4 m for the first time since NFI3 (1951–1953) (Ilvessalo 1957).

Soil type (Table 2.10) was determined at the depth of 10–30 cm from the surface. Depending on the homogeneity of the site, 2–5 observations were made, primarily at points 1–4 of Fig. 2.5. Soil type was always recorded as bedrock or stony soil if the thickness of the soil (organic and mineral layers) was less than 10 cm. If the thickness of the organic layer on top of the mineral soil was less than 30 cm or if the thickness of the soil (organic and mineral) was 10–30 cm and there was mineral soil between the organic layer and the bedrock, the soil type was determined on the basis of the mineral soil.

Mean grain size was assessed if the soil type was glacial till or sorted soil. The classes were 'fine', 'medium', and 'coarse'. *Stoniness* measurements were made only on forest land and poorly productive forest land stands and only if the thickness of the organic layer was less than 15 cm. The measurements were made at points 1–4 of Fig. 2.5 using a soil probe, and the recorded result was the average of the measurements.

On permanent plots, the type (Table 2.11) and thickness of the organic layer were assessed at points 1–4 of Fig. 2.5; on temporary plots, points 1–3 were selected. In NFI, mull soil is classified as an organic layer. *The thickness of the organic layer*

2.9 Taxation Class

Table 2.11 The types of organic layer

0	Very thin (<1 cm) or missing organic layer
1	Raw humus
2	Moder
3	Mull soil
4	Peat
5	Raw humus on peat
6	Mull-like peat

Table 2.12 Drainage situation

0	Undrained
1	Drained mineral soil forest
2	Drained peatland, drainage effect not yet visible
3	Transforming stage of drained peatland
4	Transformed stage of drained peatland

was recorded with an accuracy of 1 cm, but an accuracy of 5 cm was sufficient if the thickness was 10–30 cm and an accuracy of 10 cm if the thickness was over 30 cm. The thickness of the organic layer was measured to the depth of 4 m.

2.8 Drainage Situation

The assessments concerning the drainage situation were made from the point of view of wood production, but these variables also give information about the naturalness of the site and can be used as biodiversity indicators.

Drainage situation (Table 2.12) distinguishes undrained and drained forestry land and classifies drained peatland on the basis of the progress of the drainage effect (the drainage stage). The improvement in hydrology and lowering of the water table to increase timber productivity has been the aim of the forest drainage, and is seen also in the class definitions. Drainage for other than forestry purposes (road ditches, agricultural cut-off ditches, individual main ditches etc.) was considered only if the drainage had had an effect on tree growth or if the drainage covered the entire stand. The final aim of the drainage operation (the transformed stage) is a drained peatland site whose ground vegetation resembles one of the mineral forests site types and the hydrology does not limit the closure of the stand. The transforming and transformed sites can be classified as poorly productive forest land or unproductive land if, due to the poor fertility of the site, they cannot be classified as forest land.

2.9 Taxation Class

The income taxation of wood production was based on the potential productivity of the sites and the area owned by the land owner. All individual forest land stands were classified into site fertility classes for taxation purposes. The system was

Table 2.13 Taxation classes

0	Herb rich sites and herb rich heath sites (site fertility classes 1 and 2) on mineral soils excluding the *Pyrola* type
1	Mesic forests and herb rich heath sites of *Pyrola* type on mineral soils
2	Sub-xeric forests and mesic forests of a very thick raw humus layer
3	Xeric and barren mineral soil forests, forests of a very thick raw humus and *Pleurozium schreberi* moss layer, natural spruce swamps (productivity ≥ 1 m^3/ha per year)
4	Natural pine fens and bogs

adopted in 1921 and lasted until the end of 2005. During the years 1993–2005 it ran parallel with a revenue based income taxation system. NFI played a central role in establishing the system and provided the average production figures based on the average tree stem volume increments and timber assortment distributions by municipalities. For this purpose, the NFI field work classified forest land stands into taxation classes (Table 2.13), which closely resembled site fertility classes (Table 2.8) but also took into account possible factors that lowered the wood production capacity.

Where the productivity of forest land was significantly lower than normal for any given site fertility class it was placed into the class that best described its productivity. Examples of sites of lowered productivity are forests on bedrock or exceptionally stony soil, forests on shores susceptible to wind, forests on hills where there is frequent snow damage, paludified forests, forests on excessively wet soils and intensively burned forests. If the wood productivity of the site did not even equal that of the lowest taxation class, the site had to be classified as poorly productive forest land or unproductive land. In the inventory, the taxation class was determined based on the current stage of the site.

Drained peatland forests in the transformed stage were classified in the same way as the corresponding mineral soil forests. Drained peatlands in the transforming stage were normally placed into a higher taxation class than the corresponding undrained peatlands, but into a lower taxation class than the corresponding mineral soil forests. On undrained peatlands and on peatlands, where the drainage effect was not yet visible, it was not necessary to follow the coding of Table 2.13, if the wood productivity of the stand corresponded to another taxation class.

2.10 Retention Trees to Maintain Biodiversity of Forests

The current forest management guidelines (Metsätalouden kehittämiskeskus Tapio 1994, 2001, 2006) recommend leaving some living trees, dead trees, snags, long stumps etc. in regeneration cuttings to decay and maintain biodiversity of forests. One goal of NFI9 was to provide information about the extent of the areas with retention trees and about the volume of these trees. For this purpose, retention trees on forest land were described, either as a separate tree storey (see Sect. 2.11) if the number of these trees per hectare was high enough, or otherwise, in fact in most

2.11 Description of the Growing Stock of the Stand

Table 2.14 Retention tree classes

0	No retention trees
1	Large, saw timber-sized living retention trees
2	Pulpwood-sized living retention trees
3	Snags left on (former) regeneration area
4	Fallen trees, e.g., wind thrown retention trees
5	High stumps (minimum height 2 m) left in the regeneration cutting
6	Clusters of natural seedlings, usually under growth trees of the previous tree generation left in the regeneration cutting
7	Retention trees left as a buffer zone to protect springs, streams etc. or a group of retention trees (but too few be regarded as a stand)
8	Retention trees forming a stand, possibly within a larger management stand

cases, using variables retention trees class (Table 2.14) and abundance of retention trees. *Abundance of retention trees* was assessed in terms of the number of the stems per hectare or in terms of the connected areas per hectare.

2.11 Description of the Growing Stock of the Stand

The growing stock of a forest land stand was described for most variables by *tree storeys*, and for some variables for the entire growing stock, i.e. for the combined tree storeys. The key rules for separating tree storeys were that they must be adequately distinguishable, the difference of the mean age of the tree storeys is usually at least 40 years and the volume or stem number of growing stock for both tree storeys is sufficiently high. The tree storey categories were: dominant, standards and under storey. The standards could be seed trees, shelter trees or retention trees. The under storey could be classified as usable undergrowth, non-viable undergrowth or unstable seedling material. Note that the within tree storey variation of the age of the trees could also be higher than 40 years, e.g., in the case of naturally originated stands on poor sites. The difference could also be lower than 40 years, e.g., in the case of different species, such as young spruces under broadleaved trees, particularly on fertile sites. The difference of the mean heights of the trees of the two storeys was significant in those cases. A general rule is that the tree storeys are of different development classes (Table 2.17). However, in the case of development class advanced seedling stand, separate tree storeys could have the same development class. In such cases the upper storey was typically broadleaved trees.

Only the two most significant tree storeys were described (Table 2.15). The significance was assessed from the point of view of the development of the stand. Other possible parts of the growing stock were combined to these storeys. A stand was classified as uneven-aged when the growing stock consisted of a mixture of parts of clearly different storeys and even-aged structure could not be achieved with

Table 2.15 Number of tree storeys

0	Uneven-aged stand
1	Single-storied stand
2	Two-storied stand

cuttings or silvicultural operations. Selective thinning from above could be recommended for these stands. The description of the growing stock in uneven-aged stands was similar to the description of even-aged single-storey stands. Temporarily unstocked areas were classified as single-storey stands. Regeneration stands with seedlings of viable regeneration material or with retention trees were described as two storeys with treeless clear cut area as the first storey.

Rules to determine the positions of the storeys (Table 2.16) were as follows: the dominant tree storey was the tree storey that determined cutting and silvicultural operations for the stand. In a stand with two storeys, the upper storey was usually dominant. The lower storey was classified as dominant if it consisted of seedlings that were vital and the most suitable tree species for the site and if the upper storey consisted of seed trees or shelter trees that could already be removed (from the silvicultural point of view). The lower storey was also classified as dominant if the upper storey was mature or low-yielding and if the upper storey was not so dense that its removal would lead to the destruction of the lower storey. Development class (Table 2.17) describes the development phase of the growing stock in relation to the expected rotation. It also reflects possible cutting regimes.

The establishment type of a stand separated artificially and naturally regenerated stands, and the artificially regenerated stands further into succeeded and failed ones. It was assessed for seedling and thinning stands (development classes 2–5) and also for temporarily unstocked stands if already planted or seeded growing stock was completely damaged. A forest stand was classified as artificially regenerated if the major part of the trees capable for further development, or left in a thinning, were planted or seeded. The artificial regeneration was failed if the number of vital seedlings planted or seeded was less than the minimum number of stems per hectare required for a stand capable of further development. However, a failed artificial regeneration stand was not necessarily low-yielding if the growing stock had been complemented with natural regeneration.

Tree species proportions were described in NFI9 more thoroughly than in the previous inventories as a contribution to biodiversity monitoring. In total, 22 tree species, 7 species groups plus temporarily unstocked stands were listed (Table 2.29) whereas only 7 codes were available in NFI8. The proportions were assessed by tree storeys on the basis of the volume in development classes 4–8, and on the basis of the number of stems capable of further development in the seedling stand classes (2–3). The most important storeys, from the point of view of the further development of the stand, i.e. storeys with position 1, 2, or 5, (Table 2.16) were described in more detail than the other storeys.

The *dominant tree species and its proportion* were always assessed (except for development class 1). Determination of the dominant tree species started with the

2.11 Description of the Growing Stock of the Stand

Table 2.16 Position of tree storey

1	*Dominant storey*, not shelter trees.
2	*Over storey* was recorded when it was clearly distinguishable and when it should be considered in the management of the stand. At least some of the over storey trees must be seed or shelter trees or trees that were obviously left for further growing. This class also includes seed or shelter trees that were no longer necessary for the development of the dominant (lower) storey up to the point in time when they could be removed without damaging the dominant (lower) storey. After this time point, the trees were classified as retention trees.
3	*Retention tree storey* was recorded when it consisted of trees of merchantable size and it could not be regarded as seed or shelter trees or trees left for further growing as described in class 2. There must be at least 10–30 trees per hectare – if there were fewer trees they were not described as a storey. Instead, they were recorded as single retention trees as described in Sect. 2.10.
4	*Nurse crop* consists of broadleaved trees and should be of such density that it protects the existing or future spruce seedlings from frost. A single-storied nurse crop was described in a similar manner as the dominant storey.
5	*Under-storey capable of further development* was recorded when
	(a) the number of seedlings was larger than the minimum number of seedlings for a productive stand as presented in Appendix 10 of the field guide (Valtakunnan metsien 9. inventointi. Maastotyön ohjeet 1996–2003), or
	(b) the number of seedlings was less than the minimum but the seedlings would significantly support the regeneration of new growing stock. In these cases, a regeneration cutting or planting or seeding (in the case of an already cut regeneration area) was usually proposed for the coming 10 year period. The seedlings must be distributed so that no soil preparation or planting/seeding would be required for some parts of the stand. The seedlings were already established. The species of the under-storey seedlings must be suitable for the site and the upper storey must not be too dense to destroy the under-storey when removed. Usually under-storey seedlings were regenerating continuously.
6	*Under-storey not capable of further development* was recorded if the storey could not be managed to develop a productive growing stock for the stand. The reason can be that the species was not suitable for the site, that the storey was partly destroyed, or that the upper storey was so dense that its removal would destroy the under-storey seedlings. Under-storey not capable of further development was recorded only if the number of seedlings was higher than the minimum presented in Appendix 10 of the field guide (Valtakunnan metsien 9. inventointi. Maastotyön ohjeet 1996–2003).
7	*Non-established seedlings* showing regeneration capability of the site: the storey consisted of suitable species for the site but the density of the upper storey prohibited its development.

assessments of the proportions of the coniferous trees and broadleaved trees. In the case of equal proportions, the assessment was made on the basis of future treatments and predicted favoured tree species. The dominant tree species in the coniferous dominant storeys was the coniferous species with the highest proportion. A similar approach was used for broadleaved dominant storeys. The dominant species of the dominant storey was also the dominant tree species of the stand.

The *second species and its proportion*, a possible *third species and its proportion* as well as the *proportion of coniferous trees* in the total growing stock of the storey

Table 2.17 Development classes

1	*Temporarily unstocked regeneration stand*: a treeless area with possible retention trees and/or single trees that should be removed in clearing of the regeneration area. Small groups of seedlings may also occur on stands belonging to this class.
2	*Young seedling stand*: a stand with a dominant height of the dominant tree species less than 1.3 m.
3	*Advanced seedling stand*: a stand with a dominant height of the dominant tree species of more than 1.3 m. For a major part of the dominant trees (trees that are not removed in thinning), the diameter at breast height ($d_{1.3}$) must be less than 8 cm, and for the largest trees, less than 10 cm. The mean age (at breast height) of the dominant trees should not be more than 50 years in South Finland and not more than 120 years in North Finland.
4	*Young thinning stand*: a stand with a young growing stock at the thinning cuttings stage. The major part of cutting removal should be pulpwood-sized. The minimum mean age (at breast height) is 11 years and maximum 120 years in South Finland and 200 years in North Finland.
5	*Advanced thinning stand*: a stand with an older growing stock and a larger pole size than in young thinning stand. Saw-timber sized stems are typical and also the thinning removal typically contains saw timber. The minimum mean age (at breast height) is 20 years and maximum 140 years in South Finland and 200 years in North Finland. The development class was determined mainly on the basis of mean age if the site was too poor or the species was not suitable for the site to produce timber-sized trees.
6	*Mature stand*: a stand with a growing stock either old and/or large enough for the goal of the management of the stand to be regeneration cutting and establishment of a new stand but regeneration cutting had not yet been started. The maturity for regeneration was primarily determined with the help of the age of growing stock, and to some extent with the help of mean diameter. The stocked parts of strip cutting areas were classified into mature stands.
7	*Shelter tree stand*: a natural regeneration area, usually with 150–300 stems per hectare. The density and structure of the seeding trees allowed natural regeneration. In some cases, the density of the shelter trees may be so high that they must be removed in two stages. A shelter tree stand is usually created with a regeneration cutting (not by natural processes). The regeneration may also call for planting or seeding. The need for planting or seeding determined whether a shelter tree stand was classified as low-yielding or capable of further development.
8	*Seed tree stand*: a natural regeneration area usually with 30–150 fairly large seed trees per hectare (the minimum for birch is 10–30). A guiding maximum basal area for a seed tree stand when creating the stand was 5 m^2/ha.

were assessed for the important tree storeys. Furthermore, the proportion of coniferous stems of the total number of the stems of the storey was recorded in seedling stands. An exception to the previous rules in NFI9 was that the third species in seedling storeys was the broadleaved tree species that had the highest proportion of the total number of seedlings.

The *number of stems* was recorded only for development classes 2 and 3. The stems were counted on three circular plots with a radius of 2.30 m (with a total area of 50 m^2) to support the assessment. The centre points of these circles were the plot centre and two other points within the centre point stand 20 m away from the plot centre to the opposite directions.

2.11 Description of the Growing Stock of the Stand

The *mean diameter of trees* (cm) was assessed in development classes 4–8 on forest land. It was defined as the diameter of the basal area median tree, and normally estimated by the arithmetic mean of diameters of the tallied trees of the dominant crown storey. However, the mean diameter was also estimated in the field, and the field crew leader recorded his/her judgement on whether the tally trees give a reliable estimate of the mean diameter of the tree storey.

In seedling stands and on poorly productive forest land, the pole size was described by the *mean height of the trees* (dm). In North Finland the mean height was assessed for all development classes. For seedling stands, the mean height was the average height of the dominant and co-dominant seedlings. For development classes 4–8, the mean height was the height of the basal area median tree. For poorly productive forest land, the mean height was defined as the dominant height.

Age of growing stock was assessed on forest land and poorly productive forest land. The age for development classes 4–8 was defined as the weighted average of the ages of the trees with the volume of a tree as the weight. For development class 3 (advanced seedling stand), age was the average age of dominant and co-dominant trees capable of further development. The age of the dominant tree storey had to be measured; subjective assessment could be used for other storeys. The age was not assessed for tree storeys 6 and 7.

The total age was assessed as the sum of the age at breast height and the age of reaching breast height. The *age at breast height* (years) was measured from cores taken at 1.3 m height (later referred to as age cores) or using the annual branch whorls. If the annual rings could not be discerned without a microscope, the age was determined in the laboratory.

The *age of reaching breast height* is the age of a tree on obtaining a height of 1.3 m. If the stand was planted or sown or the trees grew from sprouts, the age of reaching breast height was determined in the field. For naturally regenerated stands, it was obtained using models with tree species, site fertility, and the length of thermal growing season as explanatory variables.

Basal area was assessed on forest land and poorly productive forest land. All living trees were included independently of tree species and diameter. Three angle count samples, completely within the stand if possible, were used as primary observations. The preferred locations of measurement points were the plot centre and the points 20 m in any cardinal direction from the plot centre. If these points did not fulfil the inclusion criteria, additional points were taken from locations close to these points. It was also possible to use half-circles. The basal area factor of the angle-count sample was 2 in South Finland and 1.5 in North Finland. *The basal area of the growing stock* (m^2/ha) was the mean of the three measurements. If the measurement points were not representative for the entire stand, the basal area for the growing stock was estimated. In a two-storied stand, the basal area included trees from both of the storeys. *The basal area of the second tree layer* in a two-storied stand was the basal area of the trees not belonging to the dominant tree layer.

2.12 Damages

NFI included damage assessment for the first time in NFI7 (1977–1984) and more complete assessments were made in NFI8. The assessments were further revised in NFI9 (Table 2.18). Some new damage descriptions were added, such as 'Abnormal flows of resin observed along the stem', 'Deformed stem, sweeps or forks in the stem' caused by former die-back or poor planting, for example, and 'Abnormal dying of branches in the lowest part of the crown', typically caused by fungi (normal competition of neighbouring trees was not recorded as this damage). In NFI8, in addition to damage description, needle loss of trees was assessed as a separate variable both for the entire stand and for the specific needle loss target trees. In NFI9, needle loss was no longer recorded as a stand level variable, but only at the tree level.

Damage was assessed on forest land, and only for tree storeys 1, 2, and 5 (see Table 2.16). Where several types of damage were observed only the most serious one was described. The variables used in damage assessment were symptom description, age of the damage, causal agent and degree of the damage. The symptom description indicated the type of symptoms that were observed, and which part of the trees was affected (Table 2.18).

Table 2.18 Symptom description

0	*No damage.*
1	*Dead standing trees*: dead trees or trees dying before the next growing season.
2	*Fallen or broken trees*: fallen trees or trees that were broken below the midpoint of the living crown. The trees could be dead or living. Also, trees leaning badly were regarded as fallen trees.
3	*Decayed standing trees.*
4	*Stem or root damage*: damage that was either on the trunks or on the roots not further than 1 m from the stem. The cause of the damage could be e.g., fungi, frost, browsing or harvesting.
5	*Resin flows*: abnormal flows of resin observed along the stem higher than 1.5 m from the stump (flow lower than this was recorded with code 4). The length of the visible flows had to be at least 30 cm.
6	*Dead or broken tops*: the main trunk was broken within the upper half of the living crown and the tree had not yet formed a new top recovering the damage.
7	*Other top damage*: die-back, deformation or other damage at the top that had not yet developed to damage on the trunks (code 8).
8	*Stem deformation*: sweeps or forks in the stem, caused. by, e.g., former die-back or poor planting.
9	*Branch damage*: several dead or broken branches within the living crown.
A	*Abnormally pruned crown (from below)*: abnormal dying of branches in the lowest part of the crown, typically caused by fungi. The normal competition of neighbouring trees was not recorded as damage.
B	*Defoliation*: fallen needles, leaves or shoots. Falling of needles or leaves caused by normal annual cycle or male flowering was not recorded as damage.
C	*Discolouration.*
D	*Multiple symptoms*: the growing stock was degenerating because of the age. Several different types of damage were observed.

2.12 Damages

Table 2.19 Causal agents

0	Unknown	
A	Abiotic/anthropogenic	
	A1	Wind
	A2	Snow
	A3	Frost
	A4	Other climatic factors
	A5	Fire
	A6	Soil factors
	A7	Logging
	A8	Air pollution (identified source)
	A9	Other human activity
B	Animals	
	B1	Voles
	B2	Moose, deer or reindeer (Cervidae)
	B3	Other vertebrates
	B4	Pine shoot beetles (*Tomicus* sp.)
	B5	Pine weevil (*Hylobius abietis*)
	B6	Pine sawflys (Diprionidae)
		(BA) Common pine sawfly (*Diprion pini*)
		(BB) European pine sawfly (*Neodiprion sertifer*)
	B7	Other defoliators
	B8	Spruce bark beetles (*Ips* sp.)
	B9	Other identified insect
	B0	Unidentified insect
C	Fungi	
	C1	Annosum root rot (*Heterobasidion annosum*)
	C2	Other decay fungi
	C3	Scleroderris canker (*Gremmeniella abietina*)
	C4	Pine twisting rust (*Melampsora pinitorqua*)
	C5	Blister rust (*Peridermium pini*)
	C6	Other rust fungi
	C7	Needle cast fungi
	C8	Other identified fungi
	C9	Unidentified fungi
D	Other factors	
	D1	Competition between plants

The age of the damage described when the damage had started and whether it was still continuing. The age classes for the onset of the damage were less than 2 years, 2–5 years, and more than 5 years.

The causal agent responsible for the damage was identified with the help of observed symptoms and other signs indicating the occurrence of the causing factor, e.g., insects, fungal fruiting bodies. In total, about 30 different causal agents were in the list of the agents (Table 2.19). They could be grouped into abiotic and biotic agents. Examples of abiotic agents are wind, snow, frost, fire and soil factors

Table 2.20 Degree of damage

0	*Slight*: does not affect the silvicultural quality of the stand (Sect. 2.13) and does not change the development class of the stand.
1	*Moderate*: lowers the silvicultural quality of the stand by one class (e.g., from good to satisfactory; see Sect. 2.13) or makes a low-yielding stand even less productive.
2	*Severe*: decreases the quality of the stand by more than one class, changes the development class into unstocked, or makes a low-yielding stand significantly less productive.
3	*Complete*: immediate regeneration required.

(e.g., drought, deficiency of nutrients, frozen soil). The biotic agents could be divided into insects, fungi or vertebrates. Human activity was recorded as the causal agent only in case the damage was not intended (e.g. not stand tending).

The degree of the damage was a stand level variable describing the cumulative effect of all the damage observed (Table 2.20). The baseline for the assessment was the state of the growing stock before the damage attack. The effect of the damage on the growth and yield, mortality and quality of timber was the main criteria for assessing the degree of damage.

During NFI9 (in 1998) a separate assessment of *the deficiency of potassium* was introduced. The deficiency of potassium was observed only for the forest land stands on peatland.

2.13 Silvicultural Quality of Stand

Field plot stands on forest land were classified on the basis of *silvicultural quality* into categories 'capable of further development' and 'low-yielding'. For low-yielding stands, the mean annual yield over the whole rotation period is so much lower than that of a managed stand on a similar site that the stand must be regenerated at a younger age than normal (according to the management schedules) (Metsätalouden kehittämiskeskus Tapio 1994). This assessment was made with reference to a managed stand with species suitable for the site, fully stocked and with a saw timber proportion (in m^3) of at least 45% for coniferous and 40% for broadleaf stands.

As a rule of thumb, the stand was considered to be low-yielding if the yield was less than 60% of that of a managed stand. Usually, the low-yielding stands should be regenerated immediately. In some cases, it would be profitable to postpone the regeneration due to the current or expected high increase in value (a notable increase in the percentage of merchantable or timber-sized stems in the near future). However, the rotation age of an low-yielding stand is always less than the "normal" rotation age for the site. Low-yielding stands can occur in all development classes. The reason for low yield may vary by development classes.

The rules for classifying a stand considered to be capable of further development on the basis of silvicultural quality were as follows. The quality was classified as good if the dominant species was suitable for the site and the stand had been managed

2.14 Accomplished and Proposed Measures

Table 2.21 Reasons for decreased quality

1	Age (the growing stock was over-aged)
2	Species composition
3	Over dense
4	Neglected management
5	Under-stocked
6	Cuttings (low density or poor technical quality caused by cutting or tending of a seedling stand)
7	Structure or spatial distribution of the growing stock (causing reduced yield)
8	Technical quality
9	Damage

according to the management schedules and the number of stems or the basal area of the dominant tree layers was adequate and trees were spatially distributed, evenly enough (Metsätalouden kehittämiskeskus Tapio 1994). For stands with dominant height less than 17–18 m, the basal area should be at least 95% of the minimum given in the recommendations. The corresponding limit was 85% if the dominant height was more than 17–18 m. For a satisfactory stand, the basal area limits were 80% and 70% and for a passable stand 70% and 60% respectively. A high percentage of tree species of low value, damages, poor technical quality, or the other reasons listed above could lower the silvicultural quality of the stand (Table 2.21). On an unstocked regeneration stand, the quality depended on the time since the cutting for artificial regeneration. The quality was good if the time was not more than 2 years, but low-yielding if the time was more than 4 years and the establishment of a new stock had not been started. In assessing the quality of a shelterwood and seed tree stand, time since the cutting for natural regeneration was not so important as in the case of artificial regeneration. Other factors were also considered, such as the accomplished establishment measures (i.e. site preparation and cleaning) and suitability for natural regeneration in a reasonable time.

2.14 Accomplished and Proposed Measures

Accomplished cutting operations were recorded on forest land and poorly productive forest land. The coding was changed during NFI9 (in 2001) in such a way that instead of one cutting, the three most recent cuttings during the past 10-year period were recorded. The proposed cutting operations, as well as accomplished and proposed silvicultural and soil preparation measures, were recorded only on forest land.

The observation period for accomplished cuttings and silvicultural operations was 10 years preceding the inventory time. The cuttings from the period 11–30 years before the inventory time point were also recorded but without the assessment of the cutting type. The proposals for cuttings and silvicultural measures were made

Table 2.22 Accomplished drainage operations

0	*No drainage* during the last 30 years
1	*Initial drainage*
2	*Cleaning of the ditches*
3	*Complementary drainage*
4	*Other than forestry drainage* (road ditches, agricultural ditches etc.)
5	*Blocking of the ditches*. Aiming to restore the peatland back to its natural state; this class was taken into use in NFI9

for the coming 10-year period. However, the need for planting or seeding was recorded in the case of proposed regeneration cutting with artificial regeneration.

The accomplished soil preparation and drainage operations (Table 2.22) were assessed from a 30-year period before the inventory time point. Only the latest operation and its time were recorded.

Prospective needs for drainage were assessed for the next 10 years. Suitability of peatland sites for timber production was assessed using the minimum effective temperature sums and site fertility class, which indicate the nutrient availability and balance. In assessing the suitability for timber production, the age, volume, technical quality and recovering capacity of the growing stock were also stressed. *Erroneous drainages* were assessed, together with their cause, which could be, e.g., a poor site for timber production or the fact that the site is technically unsuitable for drainage. If only a small part of the drainage of a larger peatland complex was erroneous, this was distinguished in the field work from a large-area erroneous drainage. *Ditch spacing* and *condition of ditches* were introduced in NFI9 to support the needed drainage operations and to assist in wood production scenarios.

It should also be born in mind that the proposed operations in NFI are based on the state of each single stand in which the assessment is made. These assessments did not take into account the sustainability of wood production. This should be remembered, particularly in the case of regeneration cuttings. The long-term sustainability of forestry is taken into account when making the long-term cutting scenarios (MELA 2010). Restrictions on wood production (e.g. forest protection) have been taken into account in all proposals (Tables A.29, A.31 and A.33).

2.15 Key Habitat Characteristics

Key habitats are local biodiversity hotspots that, according to the original definition, are rare and likely to host red-listed species (Nitare and Norén 1992). The key habitats are likely to maintain an important part of the biodiversity at local and landscape levels because these sites often have diverse flora and fauna that differ strongly from those of the surrounding areas. Many of the key habitats are small in area and within a regular stand. However, even habitats that could be classified as individual stands were found, especially on peatlands.

2.15 Key Habitat Characteristics

From a legislative point of view there are three types of key habitats recognised in Finland. First, *habitats of particular significance* are described in and protected by the Forest Act (1996). The landowner has to leave those areas untouched, or only cutting regimes which support the naturalness of the site are allowed. Only natural and semi-natural key habitats are considered in the Act. The landowner has the responsibility to be aware of the locations of these habitats. Secondly, *protected habitat types* are listed in and protected by the Nature Conservation Act (1996). It is prohibited to alter these habitats in such a way as to jeopardize their preservation, but it is not the landowner's responsibility to find the sites, and most of the habitats occur outside forests. Thirdly, *valuable habitats* are described in the Finnish Guidelines for Forestry Practices (Metsätalouden kehittämiskeskus Tapio 2006) and it is recommended that they and their characteristics are taken into account when managing forests. All three key habitat types are included in the FFCS forest certification criteria (FFCS 1002-1 2003).

In NFI9, the aim was to inventory all types of key habitats as given in the Forest Act, Nature Conservation Act and the other sources described above, regardless of their naturalness, and separately evaluate the naturalness and ecological value of the habitats. This was done in order to examine not only the most valuable sites, but also the potentially valuable sites.

In the legislation, the regional commonness of certain habitat types determines whether an individual habitat is considered as a Forest Act habitat or not. In NFI, however, this regional factor regarding the commonness of a certain habitat class or habitat characteristics was not taken into account (except for eutrophic fens that were not appointed a Forest Act status in Lapland).

A small size was another requirement for some habitats of the Forest Act. This requirement was included in the NFI field instructions only in 1999. Earlier, also large sites of herb-rich forests and oligo- and ombrotrophic mires with only sparse tree stands may have been regarded as Forest Act habitats, more precisely classes I, B, V, C, and E in Table 2.23.

Key habitat plot, a circle of 30 m radius, was established when the plot centre was on forest land, poorly productive forest land or unproductive land. The area of the key habitat patches was assessed within this plot. If a road, river or power line was defined as an individual 'stand' and intersected the 30 m circle, the habitat assessment was not extended beyond the intersection. Up to three habitats could be recorded in one plot. The existence of a key habitat in a stand could also be taken into account when assessing the restrictions on wood production (Sect. 2.4).

The normative minimum area for a key habitat was 300 m^2. However, springs, areas where groundwater surfaces without a visible spring, limestone areas and small rock formations did not have a minimum area, and the minimum area for open rock was 1,000 m^2. In practice, the habitats that could be considered to be individual stands, such as mire habitats and islands of mineral soil forest in undrained peatlands, were often presumed to be larger than the minimum 300 m^2 area.

The aim was to evaluate also those key habitats that were strongly altered by human activities. On drained peatlands at transforming or transformed stage, key habitat was recorded only if the original mire site type could be ascertained.

Table 2.23 Key habitat classes

1	Spring
2	Area where groundwater surfaces without a visible spring
3	Brook-side forest
4	Stand surrounding a small (<1 ha) pond
5	Fen or bog surrounding a small pond
6	Other small wetland area
7	Eutrophic paludified hardwood-spruce forest
8	Eutrophic birch fen or Eutrophic hardwood-spruce fen
9	Eutrophic pine fen
A	Mesotrophic hardwood-spruce mire
I	Oligotrophic hardwood-spruce and pine fens
B	Oligotrophic spruce mires and fens (not on forest land)
V	Ombrotrophic pine bogs (not on forest land)
C	*Sphagnum fuscum*-dominated bogs
D	Eutrophic and mesotrophic fens
E	Open fens and bogs (excluding *S. fuscum*-dominated)
F	Alluvial meadow. The influence of surface water can be seen in the vegetation
G	Dry mesotrophic herb-rich forests on mineral soils
H	Dry eutrophic herb-rich forests on mineral soils
J	Mesic mesotrophic herb-rich forests on mineral soils
K	Mesic eutrophic herb-rich forests on mineral soils
L	Moist mesotrophic herb-rich forests on mineral soils
M	Moist eutrophic herb-rich forests on mineral soils
N	Naturally regenerated rare hardwood forests
P	Islet (<1 ha) of mineral soil forest on undrained peatland
R	Gorge
S	Ravine
T	Precipice (>10 m high)
U	Open rock (soil layer at most patchy, only few trees)
W	Small formations of rock
X	Stonefields, boulder fields
Y	Sand fields
Z	Other rare biotope

The observed key habitat classes varied slightly by vegetation zones because of differences in the occurrence of the key habitats. Table 2.23 gives the set of key habitat classes that includes all key habitats in the entire country. The key habitats are described in more detail in the NFI Field Instructions (Valtakunnan metsien 9. inventointi. Maastotyön ohjeet 1996–2003). Note that habitats may actually be surroundings of a specific element in the forest, such as brook-side forest or area surrounding a spring.

The naturalness of a key habitat (Table 2.24) described how well the key characteristics of the habitat had been preserved. Human influence on the structural elements of the habitat, living and dead trees, vegetation characteristics and species present was considered. Silvicultural and other measures affect the nature of the habitat differently in different habitat types.

2.16 Tally Tree Measurements

Table 2.24 Naturalness of a key habitat

0	Natural
1	Signs of human activity present, but they have not altered the characteristics of the habitat
2	Signs of human activity present and they have altered the characteristics of the habitat to some extent
3	Strongly altered key habitat

Table 2.25 Measures accomplished on key habitats

0	Nature of the key habitat has not been taken into account during measures
1	Careful operations inside key habitat
2	Careful operations inside key habitat and the recommended buffer area
3	Key habitat left untouched during management measures
4	Key habitat and buffer left untouched
5	Key habitat managed for enhancing habitat value (e.g., removal of spruce in herb-rich forests and rare hardwood stands)
6	No measures on the key habitat or in the surrounding stands during past 30 years

Table 2.26 Ecological value of key habitat

0	No special management considerations.
1	The habitat is valuable and should be left unmanaged or managed carefully.
2	The habitat is very valuable and is a habitat of particular significance as described in and protected by the Finnish Forest Act.

The accomplished measures on a key habitat (Table 2.25) described the degree to which the characteristics of the habitat had been taken into account and preserved when operations were performed. Note that some valuable key habitats can be carefully managed, whereas it is recommended to leave others unmanaged in order to maintain biodiversity.

The ecological value of the key habitat (Table 2.26) is an approximate evaluation of whether or not the key habitat is sufficiently valuable that it has to be left unmanaged or managed carefully. Factors such as key habitat class, naturalness, and management history of the habitat, characteristics of the habitat and surrounding stands, and value for the landscape were considered. This variable also distinguished Forest Act habitats from less valuable habitats.

2.16 Tally Tree Measurements

The definition of a *tree* in the Finnish NFI is compatible with that of COST Action E43 (Gschwantner et al. 2009): a woody perennial of a species typically forming a single self-supporting main stem and having a definite crown. Bush-like species, such as bushy junipers (*Juniperus communis*) and willows (*Salix* spp.), are not trees. Goat willow (*Salix caprea*) and bay willow (*Salix pentandra*) can be either trees or bushes depending on their form.

The determination of the breast height and the height of a tree can be problematic on sloped or rugged terrains. For that purpose, *ground level* was defined as the level of the ground against the base of the stem, and on a slope, the point where the ground and the extension of the trunk intersect on the uphill side. Normally, the *origin point* of a tree was determined as the point where the pith meets the ground level. But where a tree grew on top of a stump or a rock, for example, the origin point was determined as point where the seed had most likely sprouted. The *breast height* was 1.3 m from the origin point. A *fork tree* is forked above breast height; if a tree was forked below breast height, each fork was measured as a separate tree.

Tally trees were selected using angle count (Bitterlich) sampling (Sect. 2.2.1). The sampling was restricted to *living trees* and *usable dead trees* with a minimum height of 1.35 m. A tree was declared as living if it had living branches and would survive until the next growing season. A dead tree was considered usable, if the wood could be used at least as fuel wood. Small dimensions or damaged timber did not rule out the dead tree to be of use. A broken tree was considered as a standing dead tree if the standing part of the tree included more than half of the original trunk volume and the branches were dead, and as a lying dead tree if the fallen part of the tree included more than half of the original trunk volume, and the fallen part had not been harvested. If more than half of the volume had been harvested, the broken tree was considered to be a stump and was not tallied (except in the re-measurement of permanent plots). A standing dead tree was considered to be usable if the standing part was usable, and a lying dead tree was usable if the fallen part of the tree was usable. A *stump* originated from felling a living tree or a standing dead tree, or when over half of the trunk volume had been removed from the fallen dead tree, even if the felled or sawn part had not been harvested.

Only the trees on forest land or poorly productive forest land were included in the NFI sample. Thus, the trees in parks, yards and on unproductive land, for example, were not measured. Nor were *shrubs* and *bushes*, woody perennials of a species typically not forming a single main stem and not having a definite crown. However, the angle count plot was also established on those plot centres that were not on forest land or poorly productive forest, but close enough to the stand boundary so that trees on those land-use classes could be included. *Sample trees* (every 5th tally tree in North Lapland, every 7th in the rest of the country) were measured more intensively (Table 2.27). Only the upper diameter, height, and damage were recorded from dead sample trees.

Each tally tree that was measured on those plots which were established as permanent in NFI8 was identified, coded and measured in a way depending on the status of the tree in NFI9 (Table 2.28), even if it was an unusable dead tree. A tree map was used to assist in finding the trees.

Coordinates of the tree were recorded in NFI9 on newly established permanent plots using the *bearing* and the *distance* (cm) from the plot centre. The bearing was measured to the pith of the tally tree at breast height and recorded using a scale

2.16 Tally Tree Measurements

Table 2.27 Tree-level variables

Measured from all tally trees:	
Tree type	(Table 2.28)
Coordinates of the tree (on permanent plots only)	
Tree species	(Table 2.29)
Diameter at breast height ($d_{1.3}$)	
Tree class	
Tree class specification	
Crown layer	
Measured from sample trees only:	
Origin type of the tree	(Table 2.30)
Upper diameter at the height of 6 m ($d_{6.0}$), for trees with a height of at least 8.1 m	
Bark thickness ($b = b_1 + b_2$)	
Lower limit of dead and dry branches	
Lower limit of green crown	
Height (h)	
The length of the broken part, if tree was broken	
Height increment over the past 5 years (i_h), only for coniferous trees, models were applied for broadleaved trees	
Height increment of the measurement year (i_{hm})	
Diameter increment over the past 5 years at a height of 1.3 m (= 2 × radius increment) (i_d)	
Age at 1.3 m	
Age from ground level to 1.3 m, using models	
Variables related to damage, similar to stand level damage	
Defoliation	
A possible change of tree class, with increased information after sample tree measurements, e.g., information based on increment and age boring	
A possible change in the specification of tree class	
The lengths of timber assortment classes	

ranging from 0 to 400. The horizontal distance was measured at breast height from the plot centre to the side of a tally tree facing the plot centre.

The *diameter over bark at breast height* (mm, at 1.3 m from the origin point of the tree) was measured perpendicular to the radius from a tree to the plot centre. The lowest diameter below breast height was measured if the tree was deformed at breast height. Breast height was redefined even if the old marks could be seen on trees on re-measured permanent plots. The diameter of the trees classified as dead trees in NFI8 was not measured but the old diameter was re-entered. Erroneous diameter measurements were corrected.

Tree class was based on the current tree volume and the current volume of the timber assortments, or, in the case of pulpwood size or smaller trees, on the expected volumes. The goal was a grouping in which trees of a same diameter have a similar current or expected saw-timber and pulpwood volume. The main criterion was the current or expected proportion of the highest quality saw-timber of the stem volume.

Table 2.28 Tree type (recorded on permanent plots established in NFI8)

V	Old tally tree. The tree was a tally tree in NFI8 and also in NFI9.
U	New tally tree, height over 1.3 m in NFI8. Within angle gauge in NFI9 due to diameter growth.
S	New tally tree, height under 1.3 m in NFI8.
T	New tally tree. New due to a reason other than growth, e.g., fallen tree, land use change, error or different inclusion criterion in NFI8.
K	Tally tree in NFI8, stump in NFI9, trunk removed between NFI8 and NFI9.
R	Tally tree in NFI8, stump in NFI9, trunk had not been removed.
N	Mistakenly measured tally tree in NFI8, tally tree in NFI9.
Z	Mistakenly measured tally tree in NFI8, not a tally tree in NFI9.
M	Tally tree in NFI8, land use class had changed and the tree did not exist in NFI9.
J	Tally tree in NFI8, land use class had changed, tree still existed in NFI9.
E	Tally tree in NFI8, tree could not be found in NFI9.
P	Tally tree in NFI8, but did not belong into the sample plot in NFI9.

Table 2.29 Tree species

1	Scots pine (*Pinus sylvestris*)
2	Norway spruce (*Picea abies*)
3	Silver birch (*Petula pendula*)
4	Downy birch (*Petula pubescens*)
5	European aspen (*Populus tremula*)
6	Grey alder (*Alnus incana*)
7	Black alder (*Alnus glutinosa*)
8	European mountain ash (*Sorbus aucuparia*)
9	Goat willow (*Salix caprea*)
A1	Lodgepole pine (*Pinus contorta*)
A2	Swiss stone pine (*Pinus cembra*)
A3	Other pine
A4	Larch (*Larix* sp.)
A5	Fir (*Abies* sp.)
A6	Other spruce or fir species
A7	*Thuja* (*Thuja* sp.)
A8	Common juniper (*Juniperus communis*)
A9	English yew (*Taxus baccata*)
A0	Other conifer
B1	Bay willow (*Salix pentandra*)
B2	Fluttering elm (*Ulmus laevis*)
B3	Wych elm (*Ulmus glabra*)
B4	Small-leaved lime (*Tilia cordata*)
B5	Poplar (*Populus* sp.)
B6	European ash (*Fraxinus excelsior*)
B7	Pedunculate oak (*Quercus robur*)
B8	Bird cherry (*Prunus padus*)
B9	Norway maple (*Acer platanoides*)
B0	Other broadleaved

2.16 Tally Tree Measurements

Table 2.30 Origin types of trees

0	Unknown
1	Natural seed originated
2	Natural sprout originated
3	Planted
4	Seeded

The tree class was also utilized in volume estimation, see Sect. 3.2.4. Five different classes were available for trees smaller than saw-timber sized trees, six classes for trees of saw-timber size, three classes for dead trees and three classes for stumps (of tally trees of NFI8 on re-measured sample plots). The saw-timber criteria employed by forest industry companies were used to classify trees into saw-timber and pulpwood. The reason for lowering the tree class was given in *tree class specification*.

The *crown layer* of a tree indicated its storey (Table 2.16), and its vertical position with respect to the other trees of the storey. The current position of a tree was used in development classes 2–6 (Table 2.17). The position with respect to the full stocking phase, i.e., before cutting, was used as criteria in development classes 1, 7 and 8, and also in under-productive stands treated by selective cuttings. Individual trees could also be located to storeys that were not included in the stand-level description of the growing stock (cf. Sect. 2.11).

The origin type of a tree (Table 2.30) can be used in growth and yield studies and as a stratification factor in volume estimation.

The *upper diameter* (cm) was measured at the height 6 m above the origin point of the tree from trees with heights of at least 8.1 m, perpendicular to the radius to the plot centre. The upper diameter was 0 for fork trees.

The *bark thickness* (mm) is measured from trees on temporary plots perpendicular to the radius of the plot and from both sides of each tree. The sum of the measurements is recorded. The bark thickness is used in the estimation of volume increment (Sect. 3.3).

The *lower limit of dead and dry branches* was assessed as the minimum height at which dead branches with a minimum thickness of 15 mm occur; individual dead branches in an otherwise branchless part of a trunk do not count. This variable was used in assessing the quality classes of saw-timber and the lower limit was the height at which the dead branches start to affect the quality classes of saw-timber. The variable was measured only for tree types and species capable of producing saw-timber. The *lower limit of the connected living crown* (dm) was measured from all living sample trees.

The *height of a tree* (dm) was measured along the stem axis from the origin point to the tree top. If a tree, or the main stem in case of a fork tree, was broken, then the *length of the broken part* was also measured and recorded. The length was estimated, if the broken part could not be found.

The *height increment over 5 years* (dm) was measured for coniferous trees using either binoculars or a measurement pole or using a combination of Vertex and binocular records (see Sects. 2.21 and 2.22). The height increments of broadleaved trees were estimated using height increment models, and a *growth space* code was recorded in the field to be used as an explanatory variable in the models. The 5-year height increment was also assessed, in addition to the growth space, for broadleaved trees with a height of less than 8.1 m. This change was adopted in 1997. The *height increment of the inventory year* was measured from living sample trees. Until 31 July, it is an incomplete increment and not included in the 5-year increment used in the volume increment calculations.

Diameter increment over 5 years at the height of 1.3 m was measured in laboratory from increment cores. The increment cores were bored to the pith in order to obtain data for age calculation and growth variation modelling. The increment borings and bark thickness measurements were carried out only on temporary plots. When it was not possible to obtain a complete increment core, e.g., from a partly decayed tree, the increment was calculated in the field (as 2 times the radial increment).

The *age at breast height* (1.3 m) of a sample tree was calculated from the increment core. In the case of a broken core, the age was either calculated from the core in the field or by counting the whorls of a tree. The *age from ground level to breast height* was assessed either by counting the whorls or by using the known or assessed cultivation time, or by recording the site fertility codes for assessing the time to reach breast height. These codes were inputs for the models to assess the time.

The damage assessments of sample trees were similar to those of stand level assessments (Sect. 2.12). The variables *symptom description*, *age of the damage*, *causal agent*, and *degree of the damage* had the same values as the respective stand level variables (Tables 2.18 and 2.19), except that the description 'multiple symptoms' was not applied at the tree level.

The *defoliation class* was assessed for coniferous trees using 5% class intervals. The target trees for defoliation assessments were standards in development classes 'young seedling stand' and 'advanced seedling stand' and trees in the dominant storey in other development classes.

With more complete information from a sample tree, e.g., through boring, it is possible to assess the tree class more precisely. The *possible change of tree class* was recorded. The original tree class remained unchanged. The *possible change in the specification of tree class* was similarly recorded.

The bucking of a sample tree was carried out using the recorded stem lengths, taper curve models, the quality and length requirements, the relative prices and optimisation of the value of the timber, see Sect. 3.2.3. The lengths of different timber assortment quality classes were recorded in the field. The trees with a saw-timber part were divided into *quality parts*, i.e., the connected parts of a stem of one timber assortment class that did not contain a mandatory cut point. The mandatory cut point was also recorded as a separate quality part. The quality class (Table 2.31), the *length* of the part (in decimetres; 0 for the cut point) and a possible *quality lowering reason* or the reason for the mandatory cut were recorded for each quality part. The quality classes and the criteria for lowering the quality were determined as a compromise of different quality requirements set by forest industries. Examples of reasons for lowering the quality or recording a mandatory cut point are branches,

2.18 Keystone Tree Species

Table 2.31 Timber assortment quality classes

1	Branch-free saw-timber (high quality saw-timber)
2	Saw-timber with living branches
3	Saw-timber with dead branches
4	Bottom part of the stem usable as pulpwood but not as saw-timber
5	A part in the middle of the stem usable as pulpwood but not as saw-timber (recorded only for broadleaved trees)
6	Waste wood (unusable for pulpwood)
7	Saw-timber part of a fork tree
8	Mandatory cut point in the middle of a quality part

sweepness, decay, and defected stem. The only length requirement for the quality part was the minimum saw-timber length in the case that parts above and below had a poorer quality.

2.17 Epiphytic Lichens

The abundance of some epiphytic lichens growing on trunks and branches of trees was assessed as an air purity indicator. The assessments were carried out on the permanent plots in development classes 4–8 with at least three coniferous tally tress with a $d_{1.3}$ of at least 5 cm. The following groups were assessed: beard lichens, (the most vulnerable lichens) (*Alectoria*, *Bryoria* and *Usnea* species), leaf lichens (e.g. *Hypogymnia*, *Parmelia* and *Pseudevernia* species, usually more resistant to air pollution than beard lichens), as well as *Scoliciosporum clorococcum* and *Desmococcus olivaeus* species.

2.18 Keystone Tree Species

Many broadleaved tree species have an important role in maintaining the biodiversity of flora and fauna. These tree species are usually rare in managed boreal forests. The sampling errors of the frequency and volume estimates of these species would be high when using sparse angle count sampling. Therefore the keystone species were measured from fixed-radius plots. The radii of the plots were 12.52 and 12.45 m in South and North Finland respectively. Individual trees of keystone species within the plot were measured from the centre point stands on either forest land or poorly productive forest land. Species-specific minimum thresholds for diameter were assigned according to biodiversity-related characteristics of the species. The location of the pith of the tree determined whether the tree belonged to the plot or not. The locations of the keystone trees were mapped on the permanent plots. The minimum diameter thresholds were:

- 30 cm for European aspen (*Populus tremula*),
- 20 cm for Grey alder (*Alnus incana*),

- 10 cm for Black alder (*Alnus glutinosa*), European mountain ash (*Sorbus aucuparia*) and Goat willow (*Salix caprea*)
- 5 cm for Fluttering elm (*Ulmus laevis*), Wych elm (*Ulmus glabra*), Small-leaved lime (*Tilia cordata*), European ash (*Fraxinus excelsior*), Pedunculate oak (*Quercus robur*), Norway maple (*Acer platanoides*) and Hazel (*Corylus avellana*).

2.19 All Tree Species

The abundances and occurrences of tree species and their temporal and spatial variations can be assessed by means of permanent plots more efficiently than by temporary plots. All tree species were identified on the permanent plots if the plot centre was on forest land, poorly productive forest land or unproductive land. The tree height had to be at least 1.35 m. Only living trees were included. The radius of the plot was 12.52 and 12.45 m in South and North Finland respectively. All tree species were recorded regardless of whether they had already been recorded as tally trees. The identified species were the same as for tally trees, 22 species and 7 species groups (Table 2.29).

2.20 Dead Wood Measurements

The amount and quality of dead and decaying wood is considered to be one of the most important indicators of biodiversity in the boreal region. The natural old-growth forests of the boreal zone have a large amount of dead wood at different stages of decay. Dead wood is an important habitat for many specialized insect, polypore, moss and liverwort species. Tree species, degree of decay and trunk diameter, for instance, influence the composition of these communities. Some species are even restricted to certain kinds of dead wood.

Only usable dead trees were measured up to NFI8, not decayed trees. The measurement of decaying and decayed trees was introduced in NFI9 to describe the biodiversity of forests. Dead wood was measured on those centre point stands that belonged either to forest land or poorly productive forest land up to the plot radius of 7 m. The plot radius was 12.52 m in the first year of NFI9 (1996) when the areas of forestry centres 8 (Keski-Suomi) and 9 (Pohjois-Savo) were measured. Furthermore, in those centres dead wood measurements were done only on every second plot.

Both standing and lying dead trees and tree parts were measured, including advanced decayed trees covered by mosses. A tree was classified as standing if the angle between the tree and the terrain normal was less than 45°. Otherwise a tree was classified as lying. The measurements differed slightly for standing trees and lying trees.

A standing tree was measured if its height was at least 1.3 m and the diameter at breast height was at least 100 mm. A lying tree was measured if the diameter at the distance of 1.3 m from the thicker end was at least 100 mm and the length at least 1.3 m.

2.20 Dead Wood Measurements

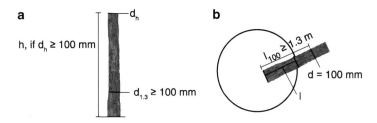

Fig. 2.6 Measurements of dead wood: (**a**) standing dead wood, (**b**) lying dead wood. The maximum radius was 12.54 m in forestry centres 8 (Keski-Suomi) and 9 (Pohjois-Savo) and 7 m elsewhere

Table 2.32 Recorded species of dead wood

0	Unidentified species
1	Scots pine (*Pinus sylvestris*)
2	Norway spruce (*Picea abies*)
3	Silver birch (*Betula pendula*)
4	Downy birch (*Betula pubescens*)
5	European aspen (*Populus tremula*)
6	Grey alder (*Alnus incana*)
7	Black alder (*Alnus glutinosa*)
8	European mountain ash (*Sorbus aucuparia*)
9	Goat willow (*Salix caprea*)
A0	Unidentified coniferous species
A1	Other coniferous species
B0	Unidentified broadleaved species
B1	Other broadleaved species
B2	Unidentified birch species

Table 2.33 Physical appearance class of dead wood

1	Unknown (usually advanced decayed lying tree)
2	Whole standing dead tree (less than 1/3 from the top broken)
3	Snag, broken standing dead tree (more than 1/3 broken)
4	Lying tree, uprooted
5	Broken tree
6	Cut stump or snag
7	Cut and abandoned butt end or bolt
8	Logging waste

Only those parts of lying trees were measured that were within the plot perimeter, met the thickness requirement, and were on forest land or poorly productive forest land (Fig. 2.6). Overall, the dead wood estimates based on NFI did not include all parts of the dead wood, for instance stem parts with a diameter of less than 100 mm, branches, stumps or roots were not included.

The common variables assessed for standing and lying trees were tree species (Table 2.32), physical appearance class (Table 2.33), *bark coverage percentage*

Table 2.34 Decay classes of dead wood (estimated by sticking a knife into the tree)

Standing trees

1. *Hard*: a knife penetrates only a few millimetres into the wood. Bark (almost) intact, branches remain. Also includes hard barkless trees, the wood of which has not started to decay.
2. *Fairly hard*: a knife penetrates 1–2 cm into the wood. Branches have started to fall off, bark of coniferous trees has started to peel off, broadleaved species often host multiple fruiting bodies of polypores on the upper part of the tree.
3. *Fairly soft*: a knife penetrates 3–5 cm into the wood. With coniferous species, bark only left at the base of the stem, with broadleaved species the bark remains but the trunk has started to decay, branches have mainly fallen off, only larger branches remain but are not intact, part of the crown has often collapsed.
4. *Soft*: a knife easily penetrates deep into the wood. The trunk remains standing only because of support from the bark, branches have totally fallen off of broadleaved species, the tree is usually broken – only the stem snag is standing.

Lying trees

1. *Hard*: a knife penetrates only a few millimetres into the wood. A recently fallen trunk with bark remaining, possible epiphytes are associated with standing trees. Also includes hard trunks that have fallen down long after dying but the wood has not started to decay.
2. *Fairly hard*: a knife penetrates 1–2 cm into the wood. The bark often remains. Only a few epiphytes that are mostly associated with standing trees.
3. *Fairly soft*: a knife penetrates 3–5 cm into the wood. The bark is often torn and peeled, locally abundant epiphytes but small growths. The category often includes e.g. pines, where sapwood is advanced decayed and only heartwood is hard.
4. *Soft*: a knife easily penetrates deep into the wood. Often barkless and covered by epiphytes. Large growths of mosses and lichens.
5. *Very soft*: disintegrates on being touched. Usually entirely covered by epiphytes, most epiphytes (mosses, lichens and shrubs) associated with forest ground, only a slight bulge distinguishes trunk from the surrounding ground.

(in classes of 20% intervals), decay class (Table 2.34), and *number of trees* (in the case of a very high number of dead tree trunks, only the mean tree was measured and the number of tree parts recorded). For standing trees, $d_{1.3}$ was measured, and also the *height* of those standing trees which were broken and the top diameter was at least 100 mm. The variables for lying trees were measured only inside the plot and were *butt diameter* (or diameter at the plot border), *top diameter* (or diameter at the plot border), *length of the tree part*, and *position to the ground,* which distinguished between tree parts not touching the ground and tree parts broken into small pieces and lying on the ground.

2.21 Equipment for Measurements

Field crews were equipped with the instruments presented in Table 2.35. Most of the measuring instruments were technically simple, but precise enough for the NFI purposes. This kind of instruments remain functional in the field conditions.

2.21 Equipment for Measurements

For instance, calliper is technically a simple tool but enables measurement of *diameter over bark at breast height* in 1 mm classes and fast. As the number of tally trees per plot is relatively small calliper is preferable instead of technically more complicated instruments. Some instruments were justified by the usefulness of the measured variable in the estimation of a derived variable, e.g. the measurement of *upper diameter* of sample trees with Finnish parabolic calliper and calliper pole. The upper diameter was measured in one cm classes only but has been proved to increase the accuracy of the stem volume estimate of the sample trees (cf. Sect. 3.2.3). A review and a study of the accuracy of some variables measured using the same instruments as in NFI9 – *diameter over bark at breast height, upper diameter, height increment of 5 years* – are presented in Päivinen et al. (1992).

The *height* of the trees was first measured with Suunto hypsometer but since 2001 with a new instrument, Vertex hypsometer. An ultrasound transponder was

Table 2.35 The equipment of a NFI9 field crew

Instrument/equipment	Producer, trademark and model
Portable computer (PC)	–
Field computer	Rufco-900 or Husky FS3
Vertex-gauge	Haglöf, Vertex III-60
Mobile phone	NOKIA 6250
GPS-receiver	Trimble Ace II
GPS-backpack	–
Binoculars	Bidox 6×30, Bresser 8×40, Optilyh 8×30
Precision compass	Suunto KB-14/400
Hypsometer	Suunto PM5-1520PCP
Increment borer, 25 and 30 cm	–
Calliper, normal size and small	–
Measuring tape, 20 and 50 m, distance marking sticks	–
Borer for stoniness	–
Compass	Suunto
Bark gauge	Suunto E60
Calliper pole, 5 m	–
Finnish parabolic Callipers	–
Relascope	–
1.3 m pole, metal or plastic	–
Earth drill	–
Peat bore, 4 m	–
Talmeter tape	–
Backpack for equipments	–
Axe, billhook and knife	–
Spray paint, permanent markers	–
First aid kit	–
Vest, raincoat, rainpants	–

used with Vertex to measure the distance from the tree. The heights can be measured at any distances from the tree using the transponder. The binoculars marked with a specific scale were used to measure the *height increment of 5 years* of conifer sample trees. The binocular records were converted to dm as a function of the distance from the tree and the height of the tree above the horizontal plane.

The data collected in the field was entered into digital form with field computers. First the Rufco-900 field computer was employed in NFI9 until 2000. After that, the Husky FS3 was used. The software necessary for data capturing was programmed in NFI using Rucfo macro language and Pascal 7.0 programming language for Husky computers.

The measurement of *diameter increment of 5 years* and the *age at breast height* (partially) from increment cores of sample trees was carried out as in-house work. First a microscope was used to measure the tree-rings. Since 2001 the majority of the ring measurements were done using WinDENDRO ring scanner and software version 6.2. WinDENDRO is a semi-automatic image analysis system specifically designed for annual tree-rings analysis (Windendro 2010).

2.22 A Correction to the Height Measurements of Year 2001

A new height measurement instrument (Vertex III) was introduced during the 2001 field season. The instrument requires a daily calibration for temperature and a further calibration that depends on the user. When analysing the early measurement data, some illogical combinations of the height and breast height diameter of the trees were noticed. Furthermore, height increments as a function of the age and diameter deviated from the earlier height increments. The functioning of the instruments was re-checked and the overestimations in both the heights and height increments were confirmed (Tomppo et al. 2003).

In order to remove systematic measurement errors, a sub-set of sample trees was re-measured in autumn 2002. The exact measurement times, both the original one and new one, were taken into account in the new measurements. The height of a tree and the height increment over 5 years as measured in the original inventory in 2001 were measured. Furthermore, either the 2-year or 1-year height increment was measured depending on whether a tree was measured in May–July or in August–September 2001. Exactly the same height points could be compared using these measurements and two different instruments. The new measurements were made using a Suunto hypsometer and a measuring tape. The height from the trees with a height less than or equal to 8 m was measured using measuring rod, although this upper limit varied to some extent between field crews. In total 1,427 of 5,085 sample trees were re-measured.

2.22.1 The Height Correction Models for the Sample Trees not Re-measured

A height correction model had to be estimated for those sample trees that were not re-measured, i.e., for the majority of the sample trees. The model was:

$$\partial h = b_i(h - c_i) + \varepsilon \qquad (2.1)$$

where

∂h is the difference of the height measurements
b_i is the parameter for field crew i,
h is the height of a tree in 2001, measured with Vertex III, minus the incomplete height increment for a tree measured in May–July 2001,
c_i is a constant within each field crew and corresponds to the height to which the height of a tree was measured using a rod, and
ε is the error term of a model distributed normally with a mean of 0.

The values of c_i by field crews and the estimates of the parameter b_i by crew i are given in Table 2.36. This type of model can be justified by the fact that its purpose is to estimate the height measurement error, not the height itself. It was assumed that the measurement error of the hypsometer was normally distributed with an expected value equal to 0. It was also assumed that the measurement error was 0 to the height that was measured using a rod. The estimated correction should also be 0 at that height in order to avoid the discontinuity of the estimated corrections.

Table 2.36 The estimates of parameter b_i by crew i and the employed values c_i of model (2.1)

Crew i	b_i	c_i (dm)
1	0.0229617076	80
2	0.0214100170	70
3	0.0202801474	50
4	0.0228972103	50
5	0.0435055007	50
6	0.0110885069	50
7	0.0225653503	50
8	0.0035038770	50
9	0.0233500157	50
10	0.0275041007	70
11	0.0317391590	70
12	0.0228483217	60

2.22.2 Models for Correcting the Height Increments

Recall the height increment measurements from Sect. 2.16. The height increment of the sample trees measured before August 1 is the increment during the 5 years preceding the inventory year, and from August 1 onwards, the increment during the measurement year and the 4 years preceding the inventory year. The increment during the measurement season is measured in both cases independently of whether or not it is complete. The models for correcting height increments had to be estimated for the sample trees without re-measurements. Separate models were derived for pine and spruce. Recall that the height increments for broadleaved trees were originally estimated using models. The height increment correction model for pine and spruce was of the form:

$$\partial i_h = b_i h + c_i i_{h,5} + \varepsilon \tag{2.2}$$

where

∂h is the difference of the height increment measurements
b_i is a parameter for the field crew i,
h is the height of a tree in 2001, measured with Vertex III, minus the incomplete height increment for a tree measured in May–July 2001,
c_i is a parameter for the field crew i
$i_{h,5}$ is the height increment measured in 2001
ε is the error term of a model distributed normally with a mean of 0.

The parameters of the model are given in Table 2.37. They could not be estimated for field crew number 8 in the case of pine when taking into account the

Table 2.37 The estimates of the parameters b_i and c_i of model (2.2) for correcting the increment measurements. The rest of the estimates of the parameters did not deviate statistically significantly from zero

Crew i	b_i	c_i
Pine		
1	−0.0025977485	0.0713221806
2	0.0050685255	0.0115224699
3	0.0085255888	0.0607187127
4	−0.0018601766	0.0585288669
5	0.0026833095	0.0878951062
6	0.0023153978	0.0473136568
7	0.0030847189	0.0200283304
9	0.0109833981	−0.0303720707
10	0.0167366709	−0.0583678684
11	0.0042402959	0.0666121481
12	−0.0077422304	0.0404369006
Spruce		
5	0.0112041026	0.0223372261
6	−0.0018758874	0.1883298238
7	−0.0142952657	0.0655153234

measurement errors. The estimates of the parameters did not deviate statistically significantly from 0. In the case of spruce, the parameters of the models could be estimated for only three field crews.

2.23 Training and Quality Assurance

One of the essential roles of a national forest inventory is to provide consistent time series data on forest resources. The assessments and measurements need to be carried out in the same manner by all inventory crews, not only during the current inventory season, but also in consecutive seasons. Only professionals are employed by the Finnish NFIs and specific training is given to ensure the consistency. A field crew in NFI9 consisted of a leader and two measurement technicians. A university degree in forestry (forester) or a degree from a university of applied sciences in forestry (forestry engineer) is a standard required from the field crew leaders. The measurement technicians also preferably have some vocational education in forestry.

From 2–4 weeks lasting training periods were organised for the crew leaders before the field season and one or two-day training sessions were arranged during the season. New crew leaders were given a longer training period than experienced ones. A one week orientation course for technical assistants concerning the tree measurements was also arranged with the crew leaders' contribution. Since many new variables were introduced in NFI9, the training was more comprehensive during the initial years than in the later ones. During NFI9 new measurement devices were also implemented, the use of which required the practice.

Comparative sessions were arranged for the field crews to control the quality of the assessments and measurements. Each field crew acted as a control crew approximately twice a season depending on the number of the crews. Separate guidelines and forms were issued for the control, comparison and reporting of the results. The control crew received the data collected by the crew subject to control and then re-measured the same sample plots and trees. The stand variables were also re-assessed. Control crew then compared the data of their own measurements to the measurements of the original crew and provided feedback of the outcomes. The results were also filed. In this way, information on possible systematic errors was obtained and recorded.

The portable computer enabled the measured data to be checked in the field. The allowed codes and the limiting values of variables were incorporated in the data recording programmes. Some variables were tested against other variables for logicality. For example, the lower limit of the green crown could not be greater than the height of a tree. In permanent plots, the measured results were compared to the former results. The most obvious errors were therefore discovered and corrected already in the field. The checking of data continued as office work after the field season. The data from different types of measurements were compiled and organised into one file with different types of records for stands, trees, dead wood, keystone tree species, and all tree species. Comparisons and checks between variables of different record types were not possible until that phase. The age and increment information from laboratory measured age cores were also combined the stand and tree data. The development class and estimated age of a stand were compared to the measured age, which also affected the silvicultural quality of a stand and the proposed measures.

An important quality assurance measure is the verification of the results. A natural mode of operation is to compare the results to those of previous inventories. All changes should be explicable, and if they are not, the reasons for them should be critically examined. The following case provides an example the importance of checking all the phases of the NFI data from forest to the final results. In 2001, a new height measurement device, the Vertex III-60, was introduced (see Sect. 2.22; Tomppo et al. 2003). The device was sensitive to temperature and also required daily calibration. Its employment was practised by the field crews. The instrument was used to measure the height of a sample tree, and also to measure the 5-year height increment of coniferous trees. In control measurements, some divergences between the field crews were detected, but they were not reported as systematic differences. The control system did not therefore respond quickly enough to the signals of odd results, but the data analysis did. In Sect. 2.22, the Vertex case is described in more detail, as well as the correction procedures. The case revealed a weak spot in the quality control system. The implementation of the device should not have occurred before a controlled experiment and stricter training sessions. The Vertex continued to be used as the main height measurement device, but with more specific instructions for its use.

In 2001, a more successful introduction of a new method occurred with the WinDENDRO semi-automatic year-ring scanner system, which was used for the age and increment core measurements (Sect. 2.21). Traditionally NFI had measured all cores using visual count and a microscope. The aim was to compare these two methods and find out if the scanner-based image analysis would be suitable for NFI purposes. In the first phase the experiment was planned measuring the increment cores of only one tree species, Scots pine. The experiment was carried out from the sampling of increment cores, measurements with both devices, analysis of the data, to the reporting of the results. When all these phases of the experiment were ready and the system revised, it was applied to increment cores of Norway spruce and birch species. The experiments proved the semi-automatic image based system to be appropriate for NFI. A spin-off of the experiment was a discovery of the effects of the different measurement techniques when using the microscope. A final result was that a routine control system for increment core measurements was implemented.

2.24 The Workload and Costs

The number of field crews per year varied from 10 to 17 during NFI9 (Table 2.38). The average number of clusters measured on forestry land was 836 per year. The number of field plots measured on land by a field crew during one day was 12.7 on

Table 2.38 The number of field crews, clusters and plots on land, length of the field work season and total working days of field crews in NFI9

Year	No. of field crews[a]	Land Clusters	Plots	Field work season Starting date	End date	Field crews Total working days	No. of field plots measured on land per day
1996	17	807	11,305	27.5.	18.10.	808	14
1997	13	974	12,253	28.5.	25.10.	803	15
1998	12	1,213	12,773	29.5.	7.11.	906	14
1999	12	1,007	10,875	1.6.	29.10.	882	12
2000	13	654	8,297	26.5.	10.10.	662	13
2001	13	718	8,998	9.6.	12.10.	848	11
2002	13	757	9,775	8.6.	4.10.	839	12
2003	11	599	6,974	5.6.	3.10.	609	11
Average	–	841	10,156	–	–	–	13
Total	–	6,729	81,250	–	–	6,357	–

[a] Some field crews worked only part of the field work season

average. On the last year only 11.6 plots on land per day were measured. This was due to the larger distances between the clusters in northern Finland, forestry centre 13 (in Fig. 2.1a), and fewer plots per cluster in the northernmost sampling density region (Fig. 2.2). The total cost of the "Monitoring of forest resources" project, which carried out the NFI9 fieldwork, data processing and report writing of the forest resources, was 8.4 million euros i.e. an average cost of 1.1 million euros per year during the 8 years of the field inventory. The budget also included the salaries of the researchers not directly involved in the field measurements. The sum consisted of 4.0 million euros salaries of the permanent staff (on average six of the crew leaders belonged to the permanent staff during NFI9) and to 4.5 million euros of other costs. The other main expenses were: the salaries of the temporary field personnel (crew leaders and field assistants), training, travelling costs of the crews, the purchase of measurement devices and other equipment, data transfer and phone bills and GIS data. The planning costs of NFI9 are not included in the above expenses.

References

Cajander AK (1909) Über Waldtypen. Acta For Fenn 1(1):1–175
Cajander AK (1926) The theory of forests types. Acta For Fenn 29:1–108
Cajander AK (1949) Metsätyypit ja niiden merkitys (Forest types and their significance). Acta For Fenn 56:1–71
Cochran WG (1977) Sampling techniques, 3rd edn. Wiley, New York
FFCS 1002-1 (2003) Criteria for group certification for the area of a forestry centre. http://www.pefc.fi/media/Standardit/FFCS_1002_1_2003ENG.pdf. Accessed 1 March 2010
Forest Act (1996) Forest Act. 1093/1996. http://www.finlex.fi/en/laki/kaannokset/1996/en19961093. Accessed 1 Mar 2010

Gschwantner T, Schadauer K, Vidal C, Lanz A, Tomppo E, di Cosmo L, Robert N, Englert Duursma D, Lawrence M (2009) Common tree definitions for National Forest Inventories in Europe. Silva Fennica 43(2):303–321

Halme M, Tomppo E (2001) Improving the accuracy of multisource forest inventory estimates by reducing plot location error – a multicriteria approach. Remote Sens Environ 78:321–327

Henttonen H (1991) VMI8:n Pohjois-Suomen otanta-asetelmien vertailu satelliittikuvatulkinnan avulla. Finnish Forest Research Institute, Helsinki, Manuscript (In Finnish)

Henttonen H (1996) Yhteenveto VMI otanta-asetelmien vertailusta satelliittikuvatulkinnan avulla. Finnish Forest Research Institute, Helsinki (In Finnish)

Henttonen H (2003) Enontekiön, Inarin ja Utsjoen otanta-asetelma (A sampling design for municipalities Enontekiö, Inari and Utsjoki). Finnish Forest Research Institute, Helsinki, Manuscript (In Finnish)

Ilvessalo Y (1957) Suomen metsät metsänhoitolautakuntien toiminta-alueittain. Valtakunnan metsien inventoinnin tuloksia. (Summary: The forests of Finland by Forestry Board Districts). Commun Inst For Fenn 47:1–128

Kalela A (1961) Waldvegetationszonen und ihre klimatischen Paralleltypen. Arch Soc Vanamo 16(Suppl):65–83

Kalela A (1970) Suomen yleisimpien metsätyyppien floristinen rakenne. In: Kalliola R (ed.) Suomen kasvimaantiede, 1973. WSOY, Helsinki, pp 261–267

Kuusela K, Salminen S (1969) The 5th national forest inventory in Finland. General design, instructions for field work and data processing. Commununicationes Instituti Forestalis Fenniae 69(4):1–72

Lehto J, Leikola M (1987) Käytännön metsätyypit. Kirjayhtymä, Helsinki, 96 p (In Finnish)

MELA ja metsälaskelmat (2010) Timber production potentials and MELA-system. http://www.metla.fi/metinfo/mela/index-en.htm. Accessed 7 Dec 2010 (In Finnish)

Metsätalouden kehittämiskeskus Tapio (1994) Luonnonläheinen metsänhoito. Metsänhoitosuositukset 1994. Metsäkeskus Tapion julkaisu 6/1994, 72 p

Metsätalouden kehittämiskeskus Tapio (2001) Metsänhoitosuositukset 2001

Metsätalouden kehittämiskeskus Tapio (2006) Hyvän metsänhoidon suositukset. Metsäkustannus, Helsinki, 100 p (In Finnish)

Nature Conservation Act (1996) Nature Conservation Act. 1096/1996. http://www.finlex.fi/en/laki/kaannokset/1996/en19961096. Accessed in Mar 2010

Nitare J, Norén M (1992) Nyckelbiotoper kartläggs i nyt projekt vid skogstyrelsen. Svensk Botanisk Tidskrift 86:219–226 (In Swedish with English abstract)

Ojansuu R, Henttonen H (1983) Kuukauden keskilämpötilan, lämpösumman ja sademäärän paikallisten arvojen johtaminen Ilmatieteen laitoksen mittaustiedoista (Derivation of local values of monthly mean temperature, effective temperature sum, and precipitation, from the observations of the Finnish Meteorological Institute). Silva Fennica 17(2):143–150 (In Finnish)

Päivinen R, Yli-Kojola H (1983) Valtakunnan metsien inventoinnin ajankäyttö. Metsäntutkimuslaitos, metsänarvioimistieteen tutkimusosasto. Manuscript. 10 p + appendices (In Finnish)

Päivinen R, Nousiainen M, Korhonen KT (1992) Puutunnusten mittaamisen luotettavuus (Summary: Accuracy of certain tree measurements). Folia Forestalia 787 (In Finnish with English summary)

Rouvinen S, Varjo J, Korhonen KT (1999) GPS-paikannuksen tarkkuus metsässä. Metsätieteen aikakauskirja 1:51–63 (In Finnish)

Tomppo E, Henttonen H, Korhonen KT, Aarnio A, Ahola A, Heikkinen J, Ihalainen A, Mikkelä H, Tonteri T, Tuomainen T (1998a) Etelä-Pohjanmaan metsäkeskuksen alueen metsävarat ja niiden kehitys 1968-97. Julkaisussa: Etelä-Pohjanmaa. Metsävarat 1968-97, hakkuumahdollisuudet 1997-2026. Metsätieteen aikakauskirja – Folia Forestalia 2B:293–374 (In Finnish)

Tomppo E, Katila M, Moilanen J, Mäkelä H, Peräsaari J (1998b) Kunnittaiset metsävaratiedot 1990–94. Metsätieteen aikakauskirja – Folia Forestalia 4B:619–839 (In Finnish)

Tomppo E, Henttonen H, Tuomainen T (2001) Valtakunnan metsien 8. inventoinnin menetelmä ja tulokset metsäkeskuksittain Pohjois-Suomessa 1992–94 sekä tulokset Etelä-Suomessa 1986–92 ja koko maassa 1986–94. Metsätieteen aikakauskirja 1B:99–248 (In Finnish)

References

Tomppo E, Tuomainen T, Henttonen H, Ihalainen A, Tonteri T (2003) Kainuun metsäkeskuksen alueen metsävarat 1969–2001. Metsätieteen aikakauskirja 2B:169–256 (In Finnish)

Tomppo E, Haakana M, Katila M, Peräsaari J (2008) Multi-source national forest inventory – methods and applications. Managing forest ecosystems, vol 18. Springer, Dordrecht, 374 p. ISBN 978-1-4020-8712-7

UNECE/FAO (2000) Forest resources of Europe, CIS, North America, Australia, Japan and New Zealand (industrialized temperate/boreal countries). UN-ECE/FAO contribution to the Global Forest Resources Assessment 2000. Main report, Timber and Forest Study Papers 17. United Nations, Geneva, 445 pp

Valtakunnan metsien 9. inventointi. Maastotyön ohjeet (1996–2003) The 9th National Forest Inventory. Field measurement instructions 1996–2003. Finnish Forest Research Institute, Helsinki

Windendro (2010) http://www.regentinstruments.com/products/dendro/DENDRO.html. Accessed 10 Nov 2010

Chapter 3
Estimation Methods

The NFI plots cover the entire land area of Finland. The main results of NFI9 presented in this report concern the areas of land use, the areas of various sub-categories of forestry land and the amount, structure and increment of the growing stock in some of those categories within forest land and poorly productive forest land. The land sub-categories are defined on the basis of ownership, site, silviculture and cutting regimes, treatments needed and growing stock, e.g., tree species composition and stand age. Area estimates of key habitats and volume estimates of dead wood are also presented.

Different types of estimators were used for the areas (Sect. 3.1), for the current growing stock (volumes, basal areas, mean diameters, and the numbers of stems; Sect. 3.2), for its increment (Sect. 3.3), and for the volume of dead wood (Sect. 3.4). Assessment of sampling errors is described in Sect. 3.5, and Sect. 3.6 presents the method used for producing thematic maps of the estimates. In the northernmost part of the country the estimation methods were modified to account for the double sampling according to Cochran (1977, chapter 12).

The results of NFI are routinely reported by forestry centre regions (Fig. 2.1a). The results presented here for South Finland, North Finland, and whole country were computed by aggregating the forestry centre level results. In those exceptional cases where a forestry centre extended to two sampling density regions (Fig. 2.1b), it was divided into two separate calculation regions. Thus, a *calculation region* is either a forestry centre or a sub-region of a forestry centre which is completely within one sampling density region. In the latter case, the forestry centre level result was computed by aggregating those of the calculation regions.

3.1 Estimation of Areas

The estimation of areas in NFI9 was based on the total land areas of the calculation regions, which are assumed to be error-free (Finland's land area by municipalities on January 1 2003), and on the numbers of sample plot centres located on the categories of interest. To be more precise, consider, as an example, the area of forest land. First, the *proportional area* of forest land within the land area of a calculation region was estimated by the corresponding proportion of sample plot centres

$$\hat{P}_{F,c} = \frac{\sum_{i \in c} y_i}{\sum_{i \in c} x_i} = \frac{n_{F,c}}{n_c}, \quad (3.1)$$

where

$\hat{P}_{F,c}$ is the estimated proportion of forest land of the total land area of calculation region c,
$i \in c$ means that the centre point of plot i is located within region c,
$y_i = 1$, when the centre point stand of the i'th sample plot was classified as forest land, and $y_i = 0$ otherwise,
$n_{F,c}$ is the number of plot centres on forest land within region c,
$x_i = 1$, when the centre point of plot i was on land and $x_i = 0$ otherwise, and
n_c is the number of plot centres on the land area within region c.

Due to the fact that the number of plot centres on land is a random variable depending on the sample, $\hat{P}_{F,c}$ is a ratio estimator (Cochran 1977). And since the inclusion probability of any stand into the inventory sample is proportional to its area, $\hat{P}_{F,c}$ is a ratio of unbiased estimators of the combined area of forest land stands in region c and of the land area of c. Such ratio estimators have a small, but usually negligible bias, when the sample size is reasonably large (Cochran 1977).

The estimator, $\hat{A}_{F,c}$, of the *absolute area* of forest land in region c was then obtained by multiplying $\hat{P}_{F,c}$ with the known land area, A_c, of the region (see Tomppo et al. 1997, 1998, 2001),

$$\hat{A}_{F,c} = \frac{n_{F,c}}{n_c} A_c, \quad (3.2)$$

and the area estimators for *aggregates of several calculation regions* are simply the sums of the area estimators of those regions. In practice, *land area represented by a sample plot centre* on land in calculation region c is defined as

$$a_c = A_c / n_c, \quad (3.3)$$

and all area estimators are sums of areas represented by the relevant plot centres.

The area estimation of key habitats was slightly different, because it was based on plot-level estimates of key habitat areas, $a_{B,i}$, rather than simple classification of the centre point stand as belonging to the habitat of interest or not (see Sect. 2.15).

3.2 Estimation of the Current Growing Stock   71

By including key habitat patches within a 30 m radius circle around the sample plot centre, even if the patches did not contain the plot centre, the inclusion probabilities of key habitats increased and more key habitats were obtained into the sample yielding a more accurate area estimate. This was considered necessary, because the key habitats often are rare, occur as small patches, or both (see Sect. 2.15).

A key habitat plot was established if the sample plot centre was on forest land, poorly productive forest land, or unproductive land; let FPU denote the combination of these classes. Within the 30 m radius plot, those key habitat patches were included in the sample, which were a part of the connected union of FPU stands containing the plot centre. The areas of those unions, $a_{FPU,i}$, were estimated for each plot i using the distances from the plot centre point to the boundaries of FPU in the four cardinal directions (if less than 30 m), yielding plot-level estimates of the proportional area of key habitat type B, $p_{B,i} = a_{B,i}/a_{FPU,i}$, for those plots with a centre point on FPU. The estimate of the proportional area of key habitat type B within the FPU land of a calculation region c was the average of the plot-level proportions

$$\hat{P}_{B,c} = \frac{1}{n_{FPU,c}} \sum_{i \in c} p_{B,i}, \qquad (3.4)$$

where n_{FPUci} is the number of plot centres on FPU within c. Finally, the estimator, $\hat{A}_{B,c}$, of the absolute area of key habitat type B within region c was the product of $\hat{P}_{B,c}$ and the usual area estimator, $\hat{A}_{FPU,c}$, of FPU land in c.

3.2 Estimation of the Current Growing Stock

The estimates of the mean and total volume and the basal area of the growing stock, numbers of stems, and the (basal area weighted) mean diameters were based on tally tree measurements (Sect. 2.16). These estimates were computed in various categories, delineated on the basis of both tree- and stand-level variables.

To illustrate the computations, let us consider the case of pine on forest land of a given calculation region c and adopt the following notations:

$N_{F,P,c}$ number of pines (P) with height greater than 1.3 m on forest land (F) within region c
d_k breast height diameter ($d_{1.3}$, cm) of the k'th pine (on forest land within region c)
$g_k = \pi(d_k/2)^2$ the basal area (cross-sectional area at breast height, cm²) of the k'th pine, assuming that the cross-section is a circle
v_k stem volume (m³) of the k'th pine
$A_{F,c}$ area of forest land (ha) within region c.

The stock parameters (for pine in the forest land of region c) considered in this section can now be defined as

$\bar{n}_{F,P,c} = N_{F,P,c}/A_{F,c}$ number of stems (per hectare)

$\bar{d}_{F,P,c} = \sum_{k=1}^{N_{F,P,c}} g_k d_k \Big/ \sum_{k=1}^{N_{F,P,c}} g_k$ basal area weighted mean diameter (cm)

$$\bar{g}_{F,P,c} = \frac{1}{A_{F,c}} \sum_{k=1}^{N_{F,P,c}} \frac{g_k}{10000} \quad \text{basal area (m}^2\text{/ha)}$$

$$\bar{v}_{F,P,c} = \frac{1}{A_{F,c}} \sum_{k=1}^{N_{F,P,c}} v_k \quad \text{mean volume (m}^3\text{/ha)}$$

$$V_{F,Pc} = \bar{v}_{F,P,c} A_{F,c} \quad \text{total volume (m}^3\text{)}$$

Parameters $\bar{n}_{F,P,c}$, $\bar{g}_{F,P,c}$, and $\bar{v}_{F,P,c}$ can be expressed in a generic form

$$\bar{y}_{F,P,c} = \frac{1}{A_{F,c}} \sum_{k=1}^{N_{F,P,c}} y_k, \tag{3.5}$$

where y_k is some characteristic of tree k; in the case of the number of stems, $y_k = 1$ for all trees k. The estimation method for stock parameters of the general form (3.5), the mean values per hectare, is first described in Sect. 3.2.1, and the estimation of the mean diameters $\bar{d}_{F,P,c}$ then described separately in Sect. 3.2.2.

The *total volumes* were estimated by the product of the estimates of mean volumes (per area unit) and the corresponding areas. The total volume estimators for *aggregates of several calculation regions* were, as in the case of areas, the sums of the total volume estimators of those regions. Finally, since the sampling density can vary between calculation regions, a mean value for an aggregate region was estimated by the ratio of the corresponding estimators of the total value and the area. In other words, each tally tree was weighted by the area represented by its sample plot centre, as defined in (3.3).

In the estimation of volumes, a further complication is that direct measurements of the tree level stem volumes were (naturally) not available. The model-based prediction of tree volumes is described in Sects. 3.2.3 and 3.2.4.

3.2.1 Mean Values per Area Unit

As discussed in Sect. 2.2.1, tree k was included as a tally tree in NFI9, if one of the sample plot centres was located within a circle, the origin of which is the location of tree k and the radius (in m) equal to

$$r_k = \min\{50 d_k / 100 \sqrt{q}, r_{\max}\}, \tag{3.6}$$

where q is the basal area factor and $r_{\max}$ the maximum tally tree plot radius (in m) employed in that region. Assuming that the cross-section of the tree at breast height is circular, i.e., has the same diameter in each direction, its inclusion probability was

$$p_k = \rho \pi r_k^2, \tag{3.7}$$

3.2 Estimation of the Current Growing Stock

where ρ is the intensity of plot centres (per m²), approximately equal to $n_c/(10,000 A_c)$. Using this approximation, the unbiased Horwitz-Thompson estimator

$$\sum_{k \in S} \frac{y_k}{p_k}, \qquad (3.8)$$

for the sum in (3.5) and the ratio estimator of Sect. 3.1 for the area of forest land $A_{F,c}$ yields the mean value estimator employed in NFI9

$$\hat{y}_{F.P.c} = \frac{n_c}{A_c n_{F,c}} \sum_{k \in S} \frac{10000 A_c y_k}{n_c \pi r_k^2} = \frac{10000}{n_{F,c}} \sum_{k \in S} \frac{y_k}{\pi r_k^2}, \qquad (3.9)$$

where $k \in S$ indicates that tree k is a pine in the forest land of calculation region c, and that it was included in the NFI9 sample.

Edge effects, discussed at length in Gregoire and Valentine (2008), were avoided in NFI9 by establishing angle count plot even if the sample plot centre was outside forest, or even outside the region, and by including all tally trees, in the case of the present example, pines on forest land the distance of which from any plot centre is equal to or less than r_k in (3.6). Thus, the inclusion probabilities follow Eq. 3.7 for all trees, regardless of their distance to stand boundaries.

The radius of the inclusion zone, r_k, was equal to r_{max}, if

$$d_k \geq d_{lim} = 100 r_{max} \sqrt{q}/50.$$

In South Finland, this threshold diameter, d_{lim}, was approximately 35.4 cm, and in North Finland, 30.5 cm. Using (3.6), we find that for trees below the threshold diameter, the terms in the sum of (3.9) are

$$u_k = q \frac{y_k}{\pi (d_k/2)^2} = q \frac{y_k}{g_k} \qquad (3.10a)$$

and for trees above it

$$u_k = \frac{y_k}{\pi r_{max}^2} = \frac{g_k}{b_{max}} \frac{y_k}{g_k}, \qquad (3.10b)$$

where b_{max} is the area (in m²) of the maximal tally tree plot, a circle with radius r_{max}.

The reason for introducing the quantities u_k is that the form heights, $fh_k = v_k/g_k$, rather than the volumes were predicted for the tally trees (Sect. 3.2.4). In terms of the predicted form heights, the estimator for the *mean volume per hectare* (m³/ha) can be written as

$$\hat{v}_{F.P} = \frac{10,000}{n_F} \sum_{k \in S} w_k fh_k, \qquad (3.11)$$

where

$$w_k = \begin{cases} q, & \text{if } d_k \leq d_{\text{lim}}, \\ g_k / b_{\text{max}}, & \text{if } d_k > d_{\text{lim}}. \end{cases} \qquad (3.12)$$

Replacing the y_k's of (3.9) by g_k's divided by 10,000 (to convert from cm² to m²), the estimator $\hat{g}_{F,P}$ of the *basal area per hectare* (m²/ha) is obtained by first counting the number of those tally trees on forest land, for which $d_k \leq d_{\text{lim}}$, multiplying the count by q, then adding to the product the sum of g_k's over the ones with $d_k > d_{\text{lim}}$ divided by b_{max}, and finally dividing by n_F. The estimator $\hat{n}_{F,P}$ of the *number of stems per hectare* is obtained by selecting $y_k = 1$, resulting in the sum of the inverse basal areas $1/g_k$ multiplied by the basal area factor plus the number of trees exceeding d_{lim} divided by b_{max}, and then multiplied by $10{,}000/n_F$.

3.2.2 Mean Diameters

The NFI9 estimator of the basal area weighted mean diameter of trees can be derived by utilizing the general result (3.9) for means per area unit. For that purpose, the true mean diameter $\overline{d}_{F,P,c}$ is expressed as a ratio of two mean values per area unit

$$\overline{d}_{F,P,c} = \sum_{k=1}^{N_{F,P,c}} g_k d_k \Bigg/ \sum_{k=1}^{N_{F,P,c}} g_k = \frac{1}{A_{F,c}} \sum_{k=1}^{N_{F,P,c}} g_k d_k \Bigg/ \frac{1}{A_{F,c}} \sum_{k=1}^{N_{F,P,c}} g_k. \qquad (3.13)$$

Applying (3.9) and (3.10a, b) to both sides of the ratio yields

$$\hat{d}_{F,P,c} = \sum_{k=1}^{N_{F,P,c}} w_k d_k \Bigg/ \sum_{k=1}^{N_{F,P,c}} w_k, \qquad (3.14)$$

with weights w_k given in (3.12). The trees below the threshold diameter, included according to the genuine angle count, all have equal weights because the sampling produces the required basal area weighting. Weighting by the basal area is needed for the large trees, which are essentially sampled from fixed radius plots.

When estimating the mean diameter over aggregates of several calculation regions, trees were further weighted by the areas represented by their sample plot centres. Thus, if $C = \{c_1, c_2, \ldots, c_p\}$ is a union of p calculation regions c_j, then the mean diameter over the trees in C was estimated by

$$\hat{d}_{F,P,C} = \sum_{j=1}^{p} \sum_{k=1}^{N_{F,P,c_j}} a_{c_j} w_k d_k \Bigg/ \sum_{j=1}^{p} \sum_{k=1}^{N_{F,P,c_j}} a_{c_j} w_k \qquad (3.15)$$

with a_{c_j}'s defined according to (3.3).

3.2.3 Predicting Sample Tree Form Factors, Volumes and Timber Assortment Proportions

Volume in the Finnish NFI is defined as tree stem volume over bark (that is, with bark), above the stump to the top of the tree, excluding branches. All trees of at least 1.35 m in height were included in the volume estimates in NFI9. Volumes and volumes by timber assortment classes were predicted for sample trees (every 7th tally tree) using volume functions and taper curve models (Laasasenaho 1976, 1982) and sample tree measurements (see Kuusela and Salminen 1969; Tomppo et al. 1997, 1998; Tomppo 2006).

The estimates of the parameters of the volume functions are available for the following tree species or species groups: pine, spruce, birch, aspen, alder, and Siberian larch (*Larix sibirica*, Ledeb.). Models for pine or birch were used for other coniferous and broad-leaved tree species respectively. The explanatory variables of the models were (measured) diameter at breast height $d_{1.3}$, (measured) upper diameter $d_{6.0}$ (for trees of a height of at least 8.1 m) and (measured) height h. Most of the models, as well as the methods to estimate the parameters of the models, are presented in Laasasenaho (1982). The exact form of the model depends on the height of a tree and on the tree species groups. Depending on the group, either the form factor or volume of a tree was the dependent variable of the model.

When the height of a tree was at least 8.1 m and all three explanatory variables $d_{1.3}$, d_6 and h were available, the parameters of the model were estimated for the form factor of a tree due to the fact that this model gave the highest precision (Laasasenaho 1982). As given in Sect. 3.2.1, the form factor is defined as $f = v/(gh)$ where v is the volume, g the breast height basal area and h the height of a tree. When the model for the form factor is multiplied to the volume, the form for pine, spruce, birch and larch with a height of at least 8.1 m is

$$v_{ob,0} = b_1 d_{1.3}^2 + b_2 d_{1.3}^2 h + b_3 d_{1.3}^3 h + b_4 d_{1.3}^2 h^2 + b_5 (d_{1.3}^2 + d_{1.3} d_{6.0} + d_{6.0}^2) + b_6 d_{6.0}^2 (h-6) \tag{3.16}$$

Laasasenaho (1982) presents the estimates of the parameters of the model (3.16) for pine, spruce and birch (two species together) (Table 3.1). The estimates for larch are given in NFI computer routines (National Forest Inventory 1996; Table 3.1).

The model for aspen (*Populus tremula* L.), grey alder (*Alnus incana* (L.) Gaertn.) and black alder (*Alnus glutinosa* (L.) Moench) is of the form of Eq. (3.17), except without the constant term for aspen

$$v_{ob,0} = a + b_1 d_{1.3}^2 + b_2 d_{1.3}^2 h + b_3 (d_{1.3}^2 + d_{1.3} d_{6.0} + d_{6.0}^2) + b_4 d_{6.0}^2 (h-6) \tag{3.17}$$

The estimates of the parameters are given in NFI computer routines (National Forest Inventory 1996; Table 3.2).

Table 3.1 The estimates of the parameters of model (3.16) by tree species groups

Species group	b_1	b_2	b_3	b_4	b_5	b_6
Pine	0.268621	−0.0145543	−0.0000478628	0.000334101	0.0973148	0.0440716
Spruce	0.208043	−0.0149567	−0.000114406	0.000436781	0.133947	0.0374599
Birch	0.226547	−0.0104691	−0.000122258	0.000438033	0.0991620	0.0334836
Larch	0.281982	−0.0152185	−0.0000980420	0.000370942	0.0914879	0.0441584

Table 3.2 The estimates of the parameters of model (3.17) by tree species groups

Species group	a	b_1	b_2	b_3	b_4
Aspen	–	0.1978357	−0.00268956	0.085609	0.0371831
Alder	0.16	0.202286	−0.0013084	0.0774162	0.0371172

Table 3.3 The estimates of the parameters of model (3.18b) by tree species groups. The bias correction, $\hat{\sigma}^2/2$, has been added to b_1

Species group	$b_1 + \hat{\sigma}^2/2$	b_2	b_3	b_4	b_5
Pine	−3.32176	2.01395	2.07025	−1.07209	−0.0032473
Spruce	−3.77543	1.91505	2.82541	−1.53547	−0.0085726
Birch	−4.49213	2.10253	3.98519	−2.65900	−0.0140970
Larch	−4.30421	2.06818	3.53310	−2.23567	−0.0111362

When only $d_{1.3}$ and h are available, i.e., the height of a tree is less than 8.1 m, and the height exceeded a tree species-specific threshold (pine 4.5 m, spruce 3.5 m, birch 6.5 m, aspen 5.0 m, alder 4.0 m and larch 1.35 m) the models for volume turned out to have higher precision than the models for the form factor or form height. The models with the highest precision for pine, spruce, birch and larch were of the form (Laasasenaho 1992)

$$v_{ob,0} = b_1 d_{1.3}^{b_2} h^{b_3} (h-1.3)^{b_4} b_5^{d_{1.3}} \tag{3.18a}$$

The parameters were estimated for the linearised model, natural logarithm of (3.18a),

$$\log(v_{ob,0}) = b_1 + b_2 \log(d_{1.3}) + b_3 \log(h) + b_4 \log(h-1.3) + b_5 d_{1.3} \tag{3.18b}$$

The parameters are presented for pine, spruce and birch in Laasasenaho (1982) and are given for larch in NFI computer routines (National Forest Inventory 1996; Table 3.3). A bias correction term, $\hat{\sigma}^2/2$, has been added to b_1, assuming the error term of the linearised model normally distributed (Table 3.3).

The model based on $d_{1.3}$ and h for aspen was

$$v_{ob,0} = b_1 d_{1.3}^{b_2} h^{b_3} (h-1.3)^{b_4} (2+1.25 d_{1.3})^{b_5} \tag{3.19}$$

3.2 Estimation of the Current Growing Stock

Table 3.4 The estimates of the parameters of model (3.19) for aspen and alder. The bias correction, $\hat{\sigma}^2/2$, has been added to b_1

Species group	$b_1+\hat{\sigma}/2$	b_2	b_3	b_4	b_5
Aspen	−4.202234	2.817006	5.535323	−3.978962	−1.048689
Alder	−3.79530	1.875970	2.950203	−1.703171	–

Table 3.5 The estimates of the parameters of model (3.20) by tree species groups. The bias correction, $\hat{\sigma}^2/2$, has been added to b_1

Species group	$b_1+\hat{\sigma}^2/2$	b_2	b_3	b_4
Pine	−4.31150	−0.009264	0.81849	2.34597
Spruce	−3.91710	0	1.04666	1.971
Birch	−4.92615	−0.010615	0.787837	2.56275
Aspen	−4.98276	−0.012403	1.047207	2.36237
Alder	−5.14953	−0.030645	0.836653	2.684341

The model for alder was the same as (3.19) except that the last term of the product was not included. The parameters for the linearised model, natural logarithm of the model (3.19) for aspen and alder are available in NFI computer routines (National Forest Inventory 1996). The estimates are shown in Table 3.4, with the bias correction, $\hat{\sigma}^2/2$, added to the constant.

Separate models of small trees were employed for trees shorter than a tree species-specific threshold, i.e. pine 4.5 m, spruce 3.5 m, birch 6.5 m, aspen 5.0 m and alder 4.0 m. Model (3.18a) and (3.18b) was used for all larch trees with $h<8.1$ m.) The form of the model for short trees was (National Forest Inventory 1996).

$$v_{ob,0} = b_1 d_{1.3}^{b_2} h^{b_3}(2+1.25d_{1.3})^{b_4} \tag{3.20}$$

The estimates of the parameters for the linearised model (natural logarithm of 3.20) with the bias correction term added to the constant are shown in Table 3.5.

3.2.3.1 Timber Assortments

The proportions of the volumes of timber assortment classes were predicted in NFI9 for sample trees using the taper curve models of Laasasenaho (1982). The models have been derived from a large data set measured in detail. The models are of the form

$$\frac{d_l}{d_{0.2h}} = b_1 x + b_2 x^2 + b_3 x^3 + b_4 x^5 + b_5 x^8 + b_6 x^{13} + b_7 x^{21} + b_8 x^{34} \tag{3.21}$$

where

$d_{0.2h}$ is the diameter at 20% height,
d_l is the diameter at a height of l from the ground level and
$x = 1 - \dfrac{l}{h}$ the relative distance from the top.

If the diameter of a tree is measured at any other point than at 20% height, the procedure by Laasasenaho (1982) can be employed to use the model. Denote the value of the ratio (3.21) at the point $x = 1 - l/h$ by t_l and the diameter at the height of l by d_l. The estimate for the diameter at the 20% height is $\hat{d}_{0.2h} = d_l / t_l$. The Eq. 3.21 and $\hat{d}_{0.2h}$ can then be used to estimate the diameters at an arbitrary height.

In NFI9, either the breast height diameter $d_{1.3}$, and the height h, or $d_{1.3}$, h and diameter at the height of 6 m, d_6, were measured for the sample trees. The basic equation (3.21) can be adjusted using the known diameters, in addition to the fact that the diameter at the height of h is zero. The difference of the quantity $z_i = d_i / \hat{d}_{0.2h}$ and the quantity t_i given by the model (3.21) at each point i, with the known diameter, can be used to construct a polynomial to pass through the known points. This correction polynomial was then added to the basic model (3.21) to get the final taper curve model.

In addition to the predicted diameters at the arbitrary height, in NFI9 account was also taken of the minimum length requirements, quality requirements and relative unit prices of the timber assortments in estimating the proportions of the volumes by timber assortment classes (saw-timber classes I–III and pulpwood). A tree stem was assumed to be cut into timber assortments in such a way as to maximize its value. In NFI9, the relative unit price classes were: saw-timber class I 3, saw-timber class II 2.5, saw-timber class III 2 and pulpwood 1. The taper curve models and the optimization procedure gave the proportions of the timber assortments of the stem volume.

3.2.4 Predicting Form Heights for Tally Trees

Form heights $fh_k = v_k / g_k$ (Sect. 3.2.1) were predicted for tally trees by the non-parametric k nearest neighbour (k-NN) estimation method. For each tally tree whose volumes were to be predicted, such sample trees are sought for which the values of the selected variables are near those of the tally tree in question. The tree level variables employed were: tree species, $d_{1.3}$, and tree class (Sect. 2.16); and the stand level variables, region code, effective temperature sum, site fertility (in terms of taxation class) and stand establishment type. A further optional variable was the distance in the geographical space.

The nearest sample trees (applicable sample trees) were selected using a kind of hierarchical approach which prioritised the variables as follows. In the first phase, only those sample trees were selected for which the values of all the categorical variables were the same as for the tally tree under consideration. The breast height diameter, $d_{1.3}$, was not allowed to deviate more than 0.5 cm from that of the tally tree. If none of the sample trees fulfilled these conditions, i.e., the set of the nearest sample trees was empty, a group of tree classes was used instead of one tree class.

3.2 Estimation of the Current Growing Stock

If the set of the nearest sample trees was still empty, $d_{1.3}$ was allowed to deviate first 1 cm, and then 2.5 cm. If the set of the nearest sample trees was still empty, the effective temperature sum was not used. The priority order of the rest of the variables was taxation class, stand establishment type, tree species (grouping was used instead of individual species) and $d_{1.3}$.

All the 'nearest sample trees' found in this way, i.e., fulfilling the condition in the search, were used in predicting the form factor of the tally tree. Let us denote the set of the nearest sample trees by S_K. The estimator for the form factor of the tally tree was obtained by replacing y_k either in Eqs. 3.10a or 3.10b, depending on the diameter, by the quantity

$$\hat{y}_k = d_{1.3,k}^2 \frac{\sum_{S_k} v_{s,k}}{\sum_{S_K} d_{1.3,s,k}^2} \qquad (3.22)$$

where

$v_{s,k}$ is the volume of sample tree s, $s \in S_K$, (used for tally tree k),
$d_{1.3,s,k}$ is its diameter, and
$d_{1.3,k}$ is the diameter of the tally tree in question and
S_K is the set of the found 'nearest' sample trees.

Note that the number of the elements in S_K, i.e., the number of the 'nearest sample trees' is not fixed.

Only the variables $d_{1.3}$ and tree species group (coniferous vs. non-coniferous) are employed for trees with $d_{1.3} < 20$ mm, due to the small number of sample trees of a small diameter.

In the case of small categories, e.g., exceptionally thick trees on poor sites, the distances from the nearest neighbours may be great, that is, similar sample trees are rare or do not exist in the current inventory for the region. A priori form height prediction was used as additional information when predicting volumes for these trees, the a priori information being the predicted volume as a function of $d_{1.3}$ by tree species group. The prediction models employed have been estimated using sample trees from neighbouring regions and/or sample trees from the previous inventory, depending on the case.

In total, 18 prediction models had been estimated for each region (6 tree species or species groups multiplied by three form height models, corresponding to total volume, saw-timber volume and waste wood volume). Sample trees from the previous inventory were not used in estimating form height models for timber assortments in NFI9, due to changes in the timber quality requirements between inventories. The weight of the a priori information was a function of the distance of the tally tree variables and the variables of the found set of the sample trees, e.g., diameters. The final form height prediction is a weighted average of the k-NN prediction and a priori prediction (Tomppo et al. 1998; Tomppo 2006).

3.3 Estimation of Volume Increment

Volume increment in the Finnish NFI means the increase in tree stem volume over bark, from above the stump to the top of the tree. The annual volume increment is calculated as an average over 5 years, based only on full growing seasons, assuming that tree growth has finished by August 1. Thus the increments in the 5 years preceding the inventory year are used for trees measured before August 1, and those in the inventory year and the 4 preceding years for trees measured on or after August 1. Let us call this period *increment measurement period* and denote its length by t.

The phases in calculating the volume increment were:

1. prediction of the volume over bark of the sample trees at the inventory time point as given in Sect. 3.2.3, called the current volume
2. prediction of the volume over bark of the sample trees in the beginning of the increment measurement period, called the past volume
3. prediction of the volume increments of the sample trees during the increment measurement period as the difference between the current volume and the past volume (Eq. 3.26)
4. calculation of the average volume increments (per year) for sample trees by diameter classes (at 1 cm intervals) and by land strata, tree species groups (pine, spruce, birch species, other broad-leaved species) and by calculation regions. The eight land strata were composed of forest land and poorly productive forest land further broken into sub-categories on the basis of mineral/peatland and drained/undrained conditions
5. calculation of the total increment (per year) for survivor trees in each stratum and in each calculation region, by multiplying the average increment for trees in each diameter class by the stem number estimate based on all tally trees in that class and summing the increments over the diameter classes
6. calculation of the final increment adding the increment of the drain to that of the survivor trees. The need for this phase is explained in Sect. 3.3.3.

The sample tree variables employed in the volume increment calculation, in addition to those required in the volume calculation, were: bark thickness, diameter increment at a height of 1.3 m (above ground) and height increment, both from the increment measurement period. The height increment was measured only for coniferous trees, while that for broad-leaved trees was predicted by means of models (Kujala 1980).

The changes in bark thickness and tree form must be taken into account in volume calculations. These changes were assessed in NFI9 using the ratio 'volume over bark divided by the basal area under bark (at a height of 1.3 m)'. It was assumed that the change in this ratio is parallel to the ratio curve calculated from a large cross-sectional set of sample trees (Kujala 1980).

3.3.1 Increment of a Sample Tree

To present the calculation of volume increments more formally, recall some notations and concepts introduced by Kujala (1980).

Denote by r the ratio volume of a tree over bark divided by basal area under bark. The ratio r may be regarded as a form height when using the basal area under bark. A model for r was estimated by tree species using the data from NFI6 (1971–1974) for Southern Finland, with over 40,000 sample trees. Denote by $\hat{r}$ the prediction of r as a function of the tree height, and its value given the height h by $\hat{r}(h)$. For a discussion of the reliability of this approach, see Kujala (1980).

Further notations are as follows:
The values of variables at the inventory time point:

1. d = diameter of tree at a height of 1.3 m at the inventory time point
2. b = double bark thickness at the inventory time point
3. h = height of tree at the inventory time point and
4. h_{-t} = height of tree at the beginning of the increment measurement period
5. $g_{ub,0}$ = basal area of tree under bark at the inventory time point $= \pi(d-b)^2/4$
6. $v_{ob,0}$ = volume of tree over bark at the inventory time point
7. $r_0 = v_{ob,0} / g_{ub,0}$ volume over bark divided by basal area under bark, both at the inventory time point
8. $\hat{r}_0 = \hat{r}(h)$ predicted ratio of current volume over bark to current basal area under bark using the model

The measured or predicted increment variables

1. i_d = diameter increment under bark in the increment measurement period
2. i_h = height increment in the increment measurement period

The variables at the inventory time point minus the increments from the increment measurement period

1. $d_{ub,-t} = d - b - i_d$ = diameter at the inventory time point under bark minus the under bark diameter increment in the increment calculation period
2. $g_{ub,-t} = \pi d^2_{ub,-t} / 4 = \pi (d - i_d - b)^2 / 4$

The following steps are taken to estimate the annual increment for a sample tree.
 1a. For a sample tree, take the current volume over bark, $v_{ob,0}$, using models 3.17–3.20 as presented in 3.2.3.
 1b. Calculate the ratio

$$r_0 = v_{ob,0} / g_{ub,0} \qquad (3.23)$$

1c. Define

$$\tilde{r}_{-t} = r_0 - (\hat{r}(h) - \hat{r}(h - i_h)) \qquad (3.24)$$

1d. Estimate the past volume over bark as

$$V_{ob,-t} = \tilde{r}_{-t} g_{ub,-t} \tag{3.25a}$$

The quantity $g_{ub,-t}$ does not exists for trees that are shorter than 1.3 m at the beginning of the increment calculation period. The volume $v_{ob,-t}$ of these trees were calculated using an approximate estimator

$$V_{ob,-t} = V_{ob,0} \frac{h_{-t}^{2.5}}{h^{2.5}} \tag{3.25b}$$

(Kujala 1980).
1e Estimate the annual increment for a sample tree as

$$i_v = (v_{ob,0} - v_{ob,-t})/t \tag{3.26}$$

It is assumed that for each individual tree, the derivative of the ratio r, dr/dh, is the same as that for the ratio predicted by the model. The bark of trees growing on poor sites is usually thicker than that of trees growing on fertile sites, which will increase the value of r. On the other hand, the form height of trees on poor sites is usually lower than that of trees on fertile sites. These facts cancel each other out to some extent, making the change in r as a function of h almost a tree species-specific constant (Kujala 1980).

Example 3.1 (Kujala 1980). For pine, $\hat{r}(h) = 0.39h + 2/(h-1.3) + 0.77\sqrt{h-1.3}$ (omitting a constant of 0.39). Let us take a pine tree with $d=16$ cm, $d_6=11$ cm, $h=12$ m, $b=15$ mm, $i_d=20$ mm and $i_h=1.9$ m. Then $g_{ub,0} = 0.016513$ m² and $g_{ub,-5} = 0.012272$ m². From the volume equations, $v_{ob,0} = 0.1207$ m³. Hence, $r_0 = 7.309$ m. From the pine model, $\hat{r}(12) = 7.386$ m, and $\hat{r}(12-1.9) = 6.450$ m. By formula (3.23) $\tilde{r}_{-5} = 6.374$, and thus by formula (3.24) $v_{ob,-5} = 6.374 \times 0.012272$ m³ $= 0.0782$ m³. Thus $i_v = 0.0085$ m³/year.

3.3.2 Increment of Survivor Trees

The increment calculation strata in NFI9 were the land use class (forest land, poorly productive forest land), main site class (mineral soil, peatland soil) and drainage situation (undrained, drained), in addition to the tree species groups (pine, spruce, birch species, other broad-leaved species). For a given stratum, the estimated increment of the survivor trees in a diameter class d is the average increment in the sample trees multiplied by the estimated number of trees in that diameter class, i.e.,

$$I_{v,l,P,d,c} = \hat{A}_{l,c}\,\hat{n}_{l,P,d,c}\, \frac{\sum_{k \in S_{s,l,P,d,c'}} i_{v,k}}{\#(S_{s,l,P,d,c'})} \tag{3.27}$$

3.3 Estimation of Volume Increment

where

$\hat{A}_{l,c}$ is the area estimate of the land stratum l in calculation region c (Eq. 3.2), e.g., undrained mineral soil forest land, F_{mu}

$\hat{n}_{l,P,d,c}$ is the estimate of the number of trees per hectare on land stratum l, in tree species group P, in diameter class d and in calculation region c, (Eq. 3.9 with $y = 1$), e.g., the estimate of the number of pine trees per hectare on land stratum F_{mu}

$S_{s,l,P,d,c'}$ stands for the set of the increment sample trees on land stratum l, in tree species group P in diameter class d and in increment calculation region c', and # means the number of the trees in the set (note that c' and c could be same but this is not requested, particularly in the case that the number of the sample trees is low), and

$i_{v,k}$ is the annual volume increment of sample tree k.

The estimated increment of survivor trees of a stratum was obtained by summing over the diameter classes

$$I_{v,l,P,c} = \sum_{d} I_{v,l,P,d,c} \tag{3.28}$$

3.3.3 Increment of Drain

Only the increments of trees that have survived until the inventory time could be measured. To calculate the total increment (per year) over the 5-year increment measurement period, the increments of the trees that have either been cut or have died naturally during the period had to be added to the increment of the survivor trees. To be more precise, consider first the sample plots measured before August 1 in a given inventory year y. The total increment 'missing' in those plots due to harvests and mortality during the increment measurement period is

$$I_{dr,y,1} = i_{dr,y-4.5} + 2i_{dr,y-3.5} + 3i_{dr,y-2.5} + 4i_{dr,y-1.5} + 5i_{dr,y-0.5}, \tag{3.29}$$

where $i_{dr,y-4.5}$ is the annual increment of the trees harvested or died between August 1 in year y–5 and the beginning of the growing season in year y–4 (all of the drain was formally allocated to the period between the growing seasons), and so on. For plots measured on or after August 1, the missing increment is

$$I_{dr,y,2} = i_{dr,y-3.5} + 2i_{dr,y-2.5} + 3i_{dr,y-1.5} + 4i_{dr,y-0.5}, \tag{3.30}$$

and the mean annual increment of the drain, missing in the increment estimate based on survivor trees only, approximately

$$\frac{1}{5} \sum_{y} \sum_{j=1}^{2} p_{y,j} I_{dr,y,j} \tag{3.31}$$

where $p_{y,1}$ is that proportion of the land area of region c which was measured before August 1 in inventory year y, $p_{y,2}$ the proportion measured from August 1 onwards in year y and the outer sum is over the inventory years y of region c.

The volumes of the drain, $v_{dr,t}$, were available for pine, spruce, and all broadleaved species by calendar year, consisting of the following components:

1. cutting removals reported by forest industry companies,
2. non-commercial roundwood removals, e.g., contract sawing and fuel wood used in dwellings,
3. estimates of harvesting losses, including those arising from silvicultural measures, based on a special study by the Finnish Forest Research Institute,
4. volume of unrecovered natural losses (2.5 million m³ per year in NFI9).

The drain for the period from August 1, year $y-1$ to the beginning of the growing season, year y was estimated by the average

$$v_{dr,y-0.5} = (v_{dr,y-1} + v_{dr,y})/2 \tag{3.32}$$

and the increment of the trees that have subsequently been cut or have died was assumed to be, on average, 70% of that of the survivor trees:

$$i_{dr,t} = 0.7 \frac{I_v}{v} v_{dr,t} \tag{3.33}$$

where I_v is the total increment and v the total volume of the survivor trees.

Combining Eqs. 3.29–3.33 yields an estimate of the mean annual increment of the drain for the increment measurement period (Salminen 1993)

$$I_{dr,c} = \frac{0.7\,I_v}{5\,v} \sum_y \sum_{j=1}^{2} p_{y,j} q_{y,j} \tag{3.34}$$

where

$$q_{y,1} = (v_{dr,y-5} + 3v_{dr,y-4} + 5v_{dr,y-3} + 7v_{dr,y-2} + 9v_{dr,y-1} + 5v_{dr,y})/2,$$

and

$$q_{y,2} = (v_{dr,y-4} + 3v_{dr,y-3} + 5v_{dr,y-2} + 7v_{dr,y-1} + 4v_{dr,y})/2.$$

3.3.4 Total Increment

Equation 3.34 was applied in each calculation region c separately to pines, spruces, and broad-leaved trees, for which separate annual drain volumes, $v_{dr,t}$, were available. The obtained increment of the drain was further divided between birch and other broad-leaved species, and between the land strata according to the corresponding proportions in the increment of survivor trees. The total increments reported in

3.5 Assessment of Sampling Error

Tables A.21a and b are the sums of the increments of survivor trees and the increments of the drain (Kuusela and Salminen 1969).

When estimating the increment of the drain in land available for wood production (Table A.21c), all harvests were attributed to the wood production land whereas natural losses were divided between wood production land and the rest of forest land and poorly productive forest land.

3.4 Estimation of the Volume of Dead Wood

The volume of dead wood was estimated on the basis of data from specifically established dead wood plots on forest land and poorly productive forest land (Sect. 2.20), circles with a fixed radius restricted to the centre point stand. The volumes of individual standing dead trees were estimated by parts of stem belonging to different diameter classes (10–20 cm, 20–30 cm, >30 cm) using the taper curve models of Laasasenaho (1982) (see also Sect. 3.2.3) with the breast height diameter and height as predictors (see Ihalainen and Mäkelä 2009 for the details). The volumes of lying dead trees were estimated using the formula for the volume of a partial cone.

Given the volume v_k of each stem or part of stem in calculation region c belonging to the dead wood stratum of interest (determined, e.g., by species in Table A.37 and additionally by the diameter class in Table A.38), the mean volume over c of dead wood in that stratum was estimated by

$$\hat{v}_{D,c} = \frac{1}{n_{FP,c}} \sum_i \left(\frac{1}{a_{FP,i}} \sum_{k \in S_i} v_k \right), \qquad (3.35)$$

where the outer sum is over those plot centres within c, which belong to forest land or poorly productive forest land, $n_{FP,c}$ is the number of such centre points, $a_{FP,i}$ is the area of that part of the centre point stand, which is inside the i'th dead wood plot, and $k \in S_i$ indicates that stem (part) k belongs to the stratum of interest and was included in the i'th dead wood plot. The total volumes and the mean volumes for aggregates of several calculation regions were estimated in the same way as in the case of the growing stock (beginning of Sect. 3.2).

3.5 Assessment of Sampling Error

For each calculation region c, the estimators of both areas (3.2) and mean volumes (3.11) can be expressed as ratio estimators

$$m_c = \frac{\sum_{i=1}^{n} y_i}{\sum_{i=1}^{n} x_i}, \qquad (3.36)$$

where i refers to a sample plot, n is the number of sample plots in the land area of region c, and y_i and x_i are appropriately selected observations from plot i. For (3.2), $y_i = A_c$, if the centre point of plot i is on forest land, otherwise $y_i = 0$, and $x_i = 1$ if the centre point of plot i is on land, otherwise $x_i = 0$. For (3.11),

$$y_i = \sum_{k \in S_i} y_k,$$

where set S_i contains the pine trees tallied from the i'th plot, and $x_i = 1$ if the centre point of plot i is on forest land, otherwise $x_i = 0$.

The method applied to estimate the variance of such ratio estimators is described in Sect. 3.5.1, and Sect. 3.5.2 explains how the variance is propagated to the combinations of such estimators (total volumes and aggregates over several calculation regions). The reported sampling errors are simply the square roots of the estimated sampling variances.

3.5.1 Sampling Error of Ratio Estimators

The basis for estimating the sampling variance of m_c (3.36) is the variation in the cluster-level residuals

$$z_r = y_r - m_c x_r,$$

where r refers to a cluster of field plots in region c,

$$y_r = \sum_{i \in r} y_i$$

and x_r is defined similarly. As explained, e.g., by Cochran (1977, section 6.3), the design-based variance can be approximated by

$$V(m_c) \approx \frac{V\left(\sum_{r \in c} z_r\right)}{\left(\sum_{r \in c} x_r\right)^2}. \tag{3.37}$$

The variance in the numerator of (3.37), in turn, is estimated by the sum of quadratic forms

$$T_g = (z_{r1(g)} - z_{r2(g)} - z_{r3(g)} + z_{r4(g)})^2 / 4, \tag{3.38}$$

over all rectangular groups g of four adjacent clusters (Fig. 3.1), yielding

$$\hat{V}(m_c) = \frac{\sum_g T_g}{\left(\sum_{r \in c} x_r\right)^2}. \tag{3.39}$$

3.5 Assessment of Sampling Error

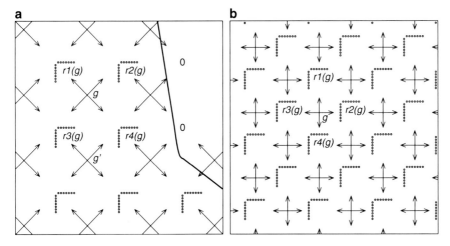

Fig. 3.1 Rectangular groups of four sample plot clusters used in estimating the sampling errors. *Left*: Groups in sampling density region 2 near the boundary with region 3. Each cluster belongs to four groups; for example, r3(g) is also r1(g') and r2(g') = r4(g). 0's indicate dummy clusters outside the region. For those, as well as for clusters r with $x_r = 0$, $z_r = 0$. *Right*: Cluster groups of the more densely sampled region 1

The *sampling error* of an area or a mean volume estimate m_c reported in the results, is the square root of the estimated variance,

$$s(m_c) = \frac{\left(\sum_g T_g\right)^{1/2}}{\sum_{r \in c} x_r}. \tag{3.40}$$

Variance estimators based on such local spatial differences were introduced into forest inventories by Matérn (1947), who refers to their earlier use in agricultural field trials (Kristensen 1933). They were further studied by Matérn (1960), and employed in Finnish and Swedish NFIs as explained by Salminen (1973) and Ranneby (1981). Similar estimators, developed for line surveys by Lindeberg (1924, 1926), were already used in the earliest NFIs (e.g., Ilvessalo 1927). A more detailed survey of their development and motivation is given by Heikkinen (2006).

For area estimators, there is a clear relationship between the area itself and the sampling error of its estimator (Fig. 3.2). This is a direct consequence of a larger number of sample plots in a larger stratum. The area estimators for North Finland are slightly less accurate than those for South Finland, due to the sparser sampling design.

Sampling error of a mean volume estimator is similarly influenced by the area of the stratum (determining, together with the sampling density, the expected number of plots within it). But it is also influenced by the density and spatial pattern of the target trees (determining, together with the employed basal area factor, the expected number of tally trees), and by the variation in the volume of individual trees within the stratum. Therefore, it is difficult to find as clear patterns as for the sampling error

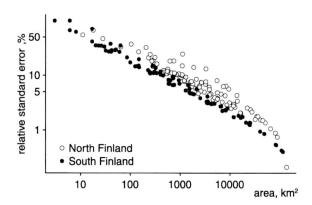

Fig. 3.2 The relationship between the estimated area, $A_{l,c}$ (Eq. 3.1), of stratum l in region c (either South Finland or North Finland), and its relative sampling error, $s(A_{l,c})/A_{l,c}$ (Eq. 3.40), for strata delineated on the basis of main site class and site fertility class (Table A.5)

of the area estimators. However, the choices made when varying the sampling density and basal area factor in NFI9 (see Sects. 2.1 and 2.2) appear to be fairly successful in the sense that the sampling errors of the mean volume are on the same level in South Finland and North Finland (see Tables A.12b and A.15a).

As an alternative to (3.39), it is quite common to "appeal to simple random sampling" when estimating the sampling error in inventory results (e.g., Scott and Gove 2002). This would mean replacing the numerator by $\sum_r z_r^2$. Both alternatives are known to yield overestimates of the variance under very general conditions (see, e.g., Heikkinen 2006, 10.5.2). However, the use of local differences can substantially decrease the overestimation, especially if there are trend-like changes over the region (e.g., Heikkinen 2006, example 10.2). For example, in South Finland the sampling errors of the area estimates of land use classes within forestry land would have been estimated to a 10–20% higher level using the simple random sampling formula (Table 3.6, Table A.1). In the more sparsely sampled North Finland, the difference is smaller, and for small and scattered classes like forest roads the two sampling error estimators yield practically identical results. Also, for the volume of growing stock the difference is over 10% in many cases (Table 3.7 and A.15a).

3.5.2 Sampling Error of Total Volumes and Aggregates

The standard approximation (e.g., Goodman 1960) based on the delta method was applied to obtain the variance of the total volume estimators:

$$V(\hat{V}_{l,P,c}) = V(\hat{v}_{l,P,c}\hat{A}_{l,c}) \approx \hat{v}_{l,P,c}^2 V(\hat{A}_{l,c}) + \hat{A}_{l,c}^2 V(\hat{v}_{l,P,c}), \qquad (3.41)$$

or, equivalently,

$$V(\hat{V}_{l,P,c})/\hat{V}_{l,P,c}^2 \approx V(\hat{A}_{l,c})/\hat{A}_{l,c}^2 + V(\hat{v}_{l,P,c})/\hat{v}_{l,P,c}^2. \qquad (3.42)$$

3.6 Thematic Maps

Table 3.6 Sampling errors of the area estimates of land-use classes within forestry land (km²) obtained using the simple random sampling formula (compare to the sampling errors reported in Table A.1)

	Forest land	Poorly productive forest land	Unproductive land	Forest roads, depots, etc.	Forestry land total
South Finland	567	140	162	52	572
North Finland	743	572	634	66	333
Whole country	935	589	654	84	662

Table 3.7 Sampling errors of the total volume of growing stock (1,000 m³) obtained using the simple random sampling formula (compare to the sampling errors reported in the 'Total' column of Table A.15a)

	Scots pine	Norway spruce	Silver birch	Downy birch	Other broadleaved	Total
South Finland	5,479	6,816	1,352	1,859	1,372	9,796
North Finland	5,949	3,875	413	2,347	554	7,169
Whole country	8,088	7,841	1,413	2,995	1,480	12,139

For aggregate regions C, containing more than one calculation region $c \in C$, the variances of the area estimators were estimated simply by

$$V(\hat{A}_{l,c}) = \sum_{c \in C} V(\hat{A}_{l,c}),$$

and those of the total volume estimators $\hat{V}_{l,P,c}$ similarly. And finally, the variance of the mean volume estimator for aggregate region C was estimated by reverting (3.41):

$$V(\hat{v}_{l,P,C}) / \hat{v}_{l,P,C}^2 \approx V(\hat{V}_{l,P,C}) / \hat{V}_{l,P,C}^2 - V(\hat{A}_{l,c}) / \hat{A}_{l,c}^2.$$

3.6 Thematic Maps

Maps of area proportions (Figs. 4.1–4.10, 4.12–4.14, 4.17, 4.18, and 4.33) and mean volumes (Figs 4.19–4.23, and 4.38) were produced from *ordinary block kriging* predictions based on cluster-level proportions of appropriate plot centre counts and cluster-level mean volumes, respectively. In order to focus on large-scale features, the target of prediction was the weighted moving average determined by a Gaussian filter with standard deviations equal to 15 km (see, e.g., Chiles and Delfiner 1999, section 3.5.2). In practice such predictions were obtained by first applying ordinary point kriging, where the degree of smoothing was based on the variogram estimated separately for each theme, and then smoothing these point kriging predictions further by the selected Gaussian filter common to all themes.

For most themes, the corresponding map was also produced from the data of NFI8, and as far as available, the corresponding maps from NFI3 were reproduced from Ilvessalo (1960).

Figure 4.15 combines predictions on three themes, area proportions of stands dominated by pine, spruce, and birch. For each map pixel it is first determined which two of these predictions are the largest, and then whether the second largest prediction is greater than 25%.

References

Chiles JP, Delfiner P (1999) Geostatistics. Wiley, New York
Cochran WG (1977) Sampling techniques, 3rd edn. Wiley, New York
Finland's land area by municipalities on January 1, 2003 (2003) Land Survey Finland. Helsinki
Goodman LA (1960) On the exact variance of products. J Am Stat Assoc 55:708–713
Gregoire TG, Valentine HT (2008) Sampling strategies for natural resources and the environment. Chapman & Hall/CRC, Boca Raton
Heikkinen J (2006) Assessment of uncertainty in spatially systematic sampling. In: Kangas A, Maltamo M (eds) Forest inventory – methodology and applications. Managing forest ecosystems, vol 10. Springer, Dordrecht, pp 155–176
Ihalainen A, Mäkelä H (2009) Kuolleen puuston määrä Etelä- ja Pohjois-Suomessa 2004–2007. Metsätieteen aikakauskirja 825:3:35–56 (In Finnish)
Ilvessalo Y (1927) The forests of Suomi Finland. Results of the general survey of the forests of the country carried out during the years 1921–1924. Communicationes ex Instituto Quaestionum Forestalium Finlandie 11 (In Finnish with English summary)
Ilvessalo Y (1960) The forests of Finland in the light of maps. Commununications Instituti Forestalis Fenniae 52(2):1–70 (In Finnish with English summary)
Kristensen RK (1933) Bestemmelse af middelfejlen ved hjælp af differensdannelser. Tidsskrift for Planteavl 39:349
Kujala M (1980) Runkopuun kuorellisen tilavuuskasvun laskentamenetelmä (Summary: A calculation method for measuring the volume growth over bark of stemwood). Folia Forestalia 441:1–8
Kuusela K, Salminen S (1969) The 5th national forest inventory in Finland. General design, instructions for field work and data processing. Commununications Instituti Forestalis Fenniae 69(4):1–72
Laasasenaho J (1976) Männyn, kuusen ja koivun kuutioimisyhtälöt. Licentiate thesis, Department of Forest Sciences, University of Helsinki, 89 p + appendices
Laasasenaho J (1982) Taper curve and volume functions for pine, spruce and birch. Seloste: Männyn, kuusen ja koivun runkokäyrä- ja tilavuusyhtälöt. Communications Instituti Forestalis Fenniae 108:1–74
Laasasenaho J (1992) Toper curve and volume functions for pine, spruce and birch. Commun Just For Fenn 108:1–74
Lindeberg JW (1924) Über die Berechnung des Mittelfehlers des Resultates einer Linientaxierung. Acta Forestalis Fennica 25:1–22
Lindeberg JW (1926) Zür Theorie der Linientaxierung. Acta Forestalis Fennica 31(6):1–9
Matérn B (1947) Methods of estimating the accuracy of line and sample plot surveys. Meddelanden från Statens Skogsforskningsinstitut 36(1) (in Swedish with English summary)
Matérn B (1960) Spatial variation. Meddelanden från Statens Skogsforskningsinstitut 49:1–144. Also appeared as Lecture Notes in Statistics 36. Springer Verlag, 1986
National Forest Inventory (1996) Volume calculation routines. Unpublished computer programs

References

Ranneby B (1981) Medelfelsformler till skattningar baserade på material från den 5:de riksskogstaxeringen. Sveriges Lantbruksuniversitet, Umeå, Institutionen för biometri och skogsindelning. Rapport 21, 19 p

Salminen S (1973) Reliability of the results from the fifth national forest inventory and a presentation of an output-mapping technique. Communicationes Instituti Forestalis Fenniae 78(6):1–64 (In Finnish with English summary)

Salminen S (1993) Eteläisimmän Suomen metsävarat 1986–1988. Folia Forestalia 825:1–111

Scott CT, Gove JH (2002) Forest inventory. In: El-Shaarawi AH, Piegorsch WW (eds) Encyclopedia of Environmetrics, vol 2. Wiley, Chichester, pp 814–820

Tomppo E (2006) The Finnish national forest inventory. In: Kangas A, Maltamo M (eds) Forest inventory – methodology and applications. Managing forest ecosystems, vol 10. Springer, Dordrecht, pp 179–194

Tomppo E, Varjo J, Korhonen K, Ahola A, Ihalainen A, Heikkinen J, Hirvelä H, Mikkelä H, Mikkola E, Salminen S, Tuomainen T (1997) Country report for Finland. In: Study on European forestry information and communication systems. Reports on forestry inventory and survey systems, vol 1, European Commission, Luxembourg, pp 145–226

Tomppo E, Henttonen H, Korhonen KT, Aarnio A, Ahola A, Heikkinen J, Ihalainen A, Mikkelä H, Tonteri T, Tuomainen T (1998) Etelä-Pohjanmaan metsäkeskuksen alueen metsävarat ja niiden kehitys 1968–97. Julkaisussa: Etelä-Pohjanmaa. Metsävarat 1968–97, hakkuumahdollisuudet 1997–2026. Metsätieteen aikakauskirja – Folia Forestalia 2B:293–374 (In Finnish)

Tomppo E, Henttonen H, Tuomainen T (2001) Valtakunnan metsien 8. inventoinnin menetelmä ja tulokset metsäkeskuksittain Pohjois-Suomessa 1992–94 sekä tulokset Etelä-Suomessa 1986–92 ja koko maassa 1986–94. Metsätieteen aikakauskirja 1B:99–248 (In Finnish)

Chapter 4
Results

4.1 The Areas of Land-Use Classes and Their Development

The estimation of area in NFI is based on the official land area of Finland as presented in Sect. 3.1. The official land area estimates from January 1st in the year succeeding the inventory of each region was used in the calculations of the NFI9 results. Finland's total land area used in the NFI9 calculations was 30.447 million ha, of which 15.461 million ha were in South Finland and 14.986 million ha in North Finland (Table 4.1, Table A.1). The land area had varied to some extent since the beginning of the 1950s due to two main reasons. The land area is still slightly increasing due to isostatic recovery following the last ice age. This is strongest in regions in Western Finland adjacent to the Gulf of Bothnia (where the adjustment is about 9 mm/year), and in South Finland along the north coast of the Baltic Sea (about 2–3 mm/year). The construction of water reservoirs has decreased the land area in North Finland by some 63,000 ha. The official land area determined by the National Land Survey (Suomen pinta-ala kunnittain 2003) has also varied during the past 50 years as earlier estimates have been replaced by more accurate ones. The change in land area resulting from the improved official data is greater that that caused by isostatic recovery and by the construction of water reservoirs.

4.1.1 Forestry Land

Forestry land consists of forest land, poorly productive forest land, unproductive land, as well as forestry roads and depots (Table 2.4). Unproductive land consists of open bogs and fens and open rocky areas or some high altitude areas in North Finland. NFI9 estimated the area of forestry land to be 26.317 million ha, of which 12.033 million ha was in South Finland and 14.284 million ha in North Finland (Table A.1, Table 4.2). Forestry land comprised 86.4% of the total land area. The area of forestry land decreased slightly in the 1950s due to clearing of arable land

Table 4.1 The areas of land classes (1,000 ha) in the whole country in the 3rd to 9th National Forest Inventories

Inventory	Field measurement years	Forest land	Poorly productive forest land	Unproductive land	Other forestry land	Forestry land	Other land	Land total
	1,000 ha							
NFI3	1951–1953	17,352	4,522	4,441	–	26,315	4,225	30,540
NFI4	1960–1963	16,909	4,832	4,492	–	26,233	4,307	30,540
NFI5	1964–1970	18,697	3,674	4,226	70	26,667	3,873	30,540
NFI6	1971–1976	19,738	3,583	3,371	86	26,778	3,772	30,550
NFI7	1977–1984	20,065	3,157	3,049	103	26,374	4,096	30,470
NFI8	1986–1994	20,074	2,983	3,093	151	26,301	4,159	30,460
NFI9	1996–2003	20,338	2,670	3,156	154	26,317	4,130	30,447

4.1 The Areas of Land-Use Classes and Their Development

Table 4.2 The areas of land classes in South and North Finland in the 8th and 9th National Forest Inventories

Inventory	Field measurement years	Forest land	Poorly productive forest land	Unproductive land	Other forestry land	Forestry land	Other land	Land total
	1,000 ha							
South Finland								
NFI8	1986–1992	11,091	541	359	82	12,074	3,393	15,467
NFI9	1996–2000	11,167	428	353	86	12,033	3,428	15,461
North Finland								
NFI8	1992–1994	8,983	2,442	2,734	68	14,227	766	14,992
NFI9	2001–2003	9,171	2,242	2,803	68	14,284	702	14,986

and built-up areas. The forest area began to increase again at the end of 1960s due to the reforestation of arable land. Since then, the construction of roads and other built-up areas has led to a slight decrease in the area of forestry land. Between NFI5 and NFI9, the area of arable land decreased by almost 700,000 ha, while the built-up area increased by 940,000 ha. Peat extraction areas, which are classified as built-up land, have decreased the forest area by 60,000 ha, and the construction of water reservoirs accounted for about 60,000 ha. On the basis of NFI9, the forestry land area was almost the same as in NFI3 (1951–1953), the difference being smaller than one standard error of the estimate (Table A.1). Note that the definitions of forestry land in the NFI3 and NFI9 differed, e.g., forest depots belonged to built up areas in NFI3.

4.1.2 Forest Land

The current definitions of forest land and of poorly productive forest land were adopted in NFI5 (1964–1970). The definition of forest land is the same as in Norway and Sweden being based on potential productivity, see Sects. 2.5 and 2.6. The definition of productive forest land in NFI1-NFI4 was based on site classes and varied to some extent. It was stricter in that sense that the production requirement was higher, and therefore the forest land area based on the earlier definition somewhat lower, than that based on the current definition. The poorest part of the current forest land was classified as poorly productive forest land (Chap. 2). Thus, the areas of forest land and the areas of poorly productive forest land area are comparable from NFI5.

The area of forest land in NFI9 was 20.34 million ha; in South Finland 11.17 million ha and in North Finland 9.17 million ha (Table 4.2, Table A.1). The area has increased by 1.64 million ha (9%) since NFI5. The main reason for the increase is peatland drainage that has converted poorly productive forest land on peatlands and also open fens and bogs to forest land. As in case of forestry land, clearing of forest land to arable land first decreased slightly forest land area since the mid of 1940s, but later, since the late 1960s and 1970s, the reforestation of arable land again increased the forest land area. Clearing forest land for built-up areas has decreased forest land area. The increase of forest land area since NFI5 is greater in North Finland than in the South. The construction of built-up areas in South Finland has therefore neutralised the benefits of peatland drainage. Furthermore, it is likely that a higher proportion of poorly productive forest land and unproductive land was converted to forest land in North Finland than in South Finland. This can be seen in the trends of respective areas (Table 4.2). Figure 4.1 shows the proportion of forest land of the land area in NFI8 and NFI9. A small increase in the forest area has taken place in East Finland and in the North-western part of South Finland where peatland soils are common and drainage has converted poorly productive forest land into forest land between NFI8 and NFI9. It should be noted that the maps in Fig. 4.1 are based on kriging prediction (Sect. 3.6), and the result depends to some extent on the locations of the observations. Some differences in the maps may also originate from

4.1 The Areas of Land-Use Classes and Their Development

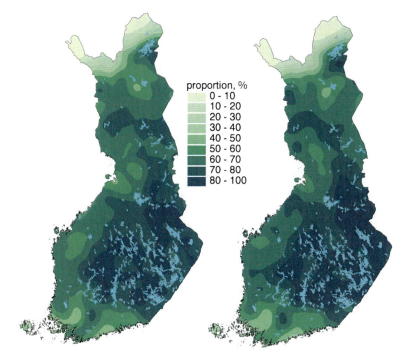

Fig. 4.1 The proportion of forest land of the land area in the 8th and 9th National Forest Inventories

the different sampling designs in NFI8 and NFI9 and do not necessarily correspond the real changes in the proportion of forest land area.

The area of poorly productive forest land has decreased continuously since NFI5 throughout the entire country, except in North Finland where it has remained approximately the same as in NFI6 and NFI5. The area of poorly productive forest land in the whole country in NFI9 was 2.67 million ha (Table 4.1, see also Fig. 4.2). Drainage of some open fens has converted a part of fens either into poorly productive forest land or forest land.

4.1.3 Land Classes Based on FAO Definitions

The land class definitions of FAO FRA 2000 and TBFRA 2000 (UNECE/FAO 2000) were also employed together with national definitions in NFI9 from 1998 onwards. The land area estimates from 1996 to 1997 were converted to correspond with the FAO land class estimates using models and other measured and assessed variables, such as site fertility class, basal area of trees and accomplished drainage treatments. The parallel use of two definitions is challenging because the assessment and measurement on a plot begins by identifying the stand or stands containing the

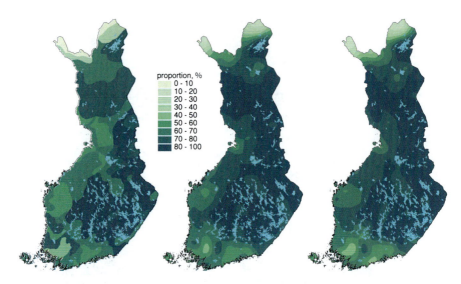

Fig. 4.2 The proportion of combined forest land and poorly productive forest land of the land area in the 3rd, 8th and 9th National Forest Inventories

field plot, or its parts, and allocating the trees on the plot to stands. The land class is a crucial variable in the stand delineation. Another factor is the different minimum size of stand in the FAO definitions and in the national definition in South Finland (0.5 and 0.25 ha respectively). Compromises were therefore necessary in some cases, in particular, in the stand delineation when adjusting the identified stands to fulfil both criteria, e.g., in cases when a part of a plot is considered as an FAO forest but not as a national forest. The area estimate based on the FAO forest definition was 22.487 million ha in NFI9, the area estimate of "other wooded land" was 0.825 million ha and the estimate of "other land" 7.136 million ha (Table A.1). "Other land" includes the poorest part of the national poorly productive forest land, nearly the entire unproductive land (a small part of unproductive land in high land areas belongs to "other wooded land"), as well as all land areas not belonging to national forestry land. The difference of the area estimate based the national forest land definition and the FAO forest definition is greater in North Finland than in South Finland due to the fact that the area of poorly productive forest land is larger in the north than in the south. The FAO forest definitions are used in international reporting, e.g. reporting for FAO FRA, MCPFE and UNFCCC LULUCF.

4.1.4 Land Use-Changes Based on the Observations on the Plot

The land use-change estimates presented above also include the effect of sampling errors on the change estimates due to the fact that temporary plots have been used

4.1 The Areas of Land-Use Classes and Their Development 99

in the inventories and estimations. The effect of the temporary plots could be reduced to some extent by using permanent plots. The permanent plots were started to be established when NFI8 proceeded to North Finland in 1992. The exclusive use of temporary field plots in the earlier inventories prevents the estimation of all land-use changes longer periods. Land-use changes have nevertheless been identified in previous inventories (Chap. 2). This information to some extent corresponds to the use of permanent plots, and can be used to assess the changes between the land-use classes. The most recent possible changes of land-use classes within the forestry land class, and the time since the changes, have been recorded during the past 10 years. Similar changes from forestry land classes into other land classes have also been recorded during the past 10 years. The changes from other land-use classes into forestry land have been recorded during the past 30 years (see Chap. 2). A total of 64,000 ha poorly productive forest land and 99,000 ha other land has converted to forest land in the entire country during the 10 years period preceding the inventory time point (Table A.2). The conversion of poorly productive forest land into forest land has occurred mainly in North Finland (46,000 ha) and it has been the result of peatland drainage. The afforestation or reforestation of arable land has converted 99,000 ha other land to forest land. This conversion has taken place mainly in South Finland (80,000 ha). An area of 121,000 ha of forest land has been lost, mainly as roads and other built-up areas. This change has occurred mainly in South Finland (110,000 ha). The observed changes on field plots can be used, e.g., for UNFCCC LULUCF reporting and in reporting under the Kyoto Protocol.

4.1.5 Ownership Information

The ownership information was obtained for each centre point stand, and also where tally trees had been measured on bi-stands. The information was obtained from the Population Register Centre of Finland. The purpose was to provide the forest resource information by ownership groups. The used ownership categories (see Chap. 2, Table 2.1) were classified into four groups for the calculation of results (Table A.3). Private persons owned 60% of the forest land area in the entire country; 75% in South Finland and 43% in North Finland. Note that the shares of the volume and volume increment of the growing stock of forests owned by the private persons are higher than the share of forest land due to the fact that the average volume and volume increment are higher in South Finland than in North Finland (e.g. Table A.15a and A.21a). The second biggest owner of the forest land area in Finland was the state; 26% in the entire country, 7% and 48% in the South and North respectively. Companies, mainly forest industry companies, owned more forests in the southern part of the country than in the northern part. The group 'Communities' includes municipalities, parishes and also jointly owned forests. Private persons owned a smaller proportion of the entire forestry land (52%) than of the forest land due to the fact that the state owns large areas in North Finland where the proportions of poorly productive and unproductive land were higher than in South Finland.

4.2 Restrictions on Forestry and Area Available for Wood Production

Because of the increased external use of NFI results, it was necessary to amend the entered data in NFI9 regarding the protected areas and other restrictions in forestry land in order to achieve a comprehensive and flexible classification. Areas classified as forestry land could in fact have more than one function, such as nature conservation, agriculture, recreation or military use. Some of the functions prevented all forest management activities, some hindered them, and some had destructive impacts on forests. The type of a protected area varied from state-owned strict nature reserves to nature reserves on private land where there were fewer restrictions on management. Land-use planning also affected forest management practices; areas were either reserved for forestry or a permit was needed for cuttings. The type of an area in itself was not considered adequate, and knowledge of the intensity of the restrictions was also required in order to present a full picture of its effect on forestry.

Six main types of restrictions and eight intensity categories were identified, see Sect. 2.4. Under each main category several sub-categories were included. One of the categories was reserved for field observations about stands with scenic values, stands with habitats of special importance, stands with effects of other land-use, and stands with constricting effects on timber harvesting. The restriction of that category was not based on any official documentation like the others, but more on a subjective judgement of the field crew leader. That information was used, for example, to estimate the alternative timber production possibilities for Forestry Centre regions (Hirvelä et al. 1998).

In Table A.4, forestry land is presented in five categories: (1) nature conservation, (2) outdoor recreation, (3) areas under land-use plan, (4) areas reserved for military use, and (5) other. The "other" category consisted of different types of areas such as antiquities, forest tree breeding and research areas. In South Finland, about 0.9 million ha of forestry land were affected by prohibitions and constrains for wood production, whereas in North Finland the area was 3.6 million ha. The high proportion in the North is due to the large amount of nature conservation areas located there. Land-use planning also set aside land to conserve valuable natural or cultural features or landscapes. Such measure could protect forests, or set limitations on forest management practices, such as needed permits for cuttings, forestation and drainage.

Some of the NFI9 results are presented only for the areas under wood production, for example proposed cuttings and silvicultural measures. The areas under wood production are given in Table 4.3. The area in South Finland was 99%

Table 4.3 The areas in wood production in South and North Finland in the 9th National Forest Inventory

	Forest land	Poorly productive forest land	Total
	1,000 ha		
South Finland	11,038	390	11,428
North Finland	8,166	1,519	9,685
Whole country	19,204	1,909	21,113

and in North Finland 85% of the total forest land and poorly productive forest land area. Areas in which all forest management activities were prohibited, were excluded from the wood production area.

4.3 Soil Classification and the Areas of Site Fertility Classes on Mineral Soils

Soil classification dates back to the early 1900s (e.g., Glinka 1914) and was developed nationally mainly for agricultural purposes. As a result, there are many different classification systems in the world, but there are also international soil classification systems (FAO-UNESCO 1988; World Reference Base for Soil Resources 1998; IUSS working group WRB 2006). The main formation process, podzolization of Finnish forest soils has been studied, e.g., by Aaltonen (1935, 1939, 1941, 1947, 1951).

Recall the soil type classification of NFI9 (Chap. 2). The classes are *organic, solid rock* and *stony soil, boulder field, glacial till* and *sorted soil*. The areas of the mineral soil types are given in Table A.6 by site fertility classes for forest land. The areas of glacial till and sorted soil are also given by the mean particle size.

The most common mineral forest soil in Finland is glacial till. The proportion of forest land on mineral soils was estimated by NFI9 to be 72% in South Finland and 81% in North Finland (Table A.6). The medium coarse category was the most common mean particle size. The proportions of the medium coarse glacial till of the forest land area on mineral soils were 63% and 75% for South and North Finland respectively. Sorted soil was the second commonest mineral soil class being, 23% of the area of mineral soil forests in South Finland and 17% in North Finland.

Some 92% of the poorly productive forest land and 77% of the unproductive land on mineral soils in South Finland was on solid rock (bedrock). In North Finland, the commonest soil of the poorly productive forest land and unproductive land on mineral soils was again glacial till, 69% and 79%, respectively. The reason for the considerable difference between South and North Finland could be the fact that all glacial till soils meet the production requirements of forest land in the southern part of the country with better growing conditions (a higher effective temperature sum during the growing season), which is not the case in the northern part of the country, particularly in the northernmost Finland.

Tamminen and Tomppo (2008) present a method to convert soil classes to the WRB system using the variables recorded in NFI9. First, fine-textured and other soils were separated using logistic regression analysis, with NFI soil variables, effective temperature and elevation as predictors. The fine-textured soils were further divided into three groups, Cambisols, Gleysols and Regosols, using discriminant analysis. After the fine-textured soils were separated, Podzols and Arenosols were separated from each other again using logistic regression analysis. Podzols were classified into the subgoups of Carbic (Carbic + Ortsteinic + Gleyic), Entic and Haplic by means of discriminant analysis. The Arenosols were also divided into three groups, Brunic, Gleyic and Haplic (Albic + Haplic).

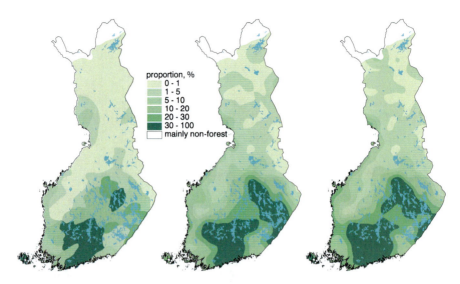

Fig. 4.3 The proportion of herb-rich forests and herb-rich heath forests of the area of the forest land in in the 3rd, 8th and 9th National Forest Inventories

The most frequent soil types in the NFI9 data were Podzols (50%), Histosols (25%), Arenosols (11%) and Leptosols (9%). Finer-textured soils formed only a small proportion: Cambisols (1.9%), Gleysols (1.4%) and Regosols (1.2%). In the WRB system, the most of the glacial till belongs to the (Haplic) Podzols category (Tamminen and Tomppo 2008).

The areas of site fertility classes and their sampling errors are given separately for mineral soils (Upland forests) and different types of peatland forests (spruce mires, pine mires and open fens and bogs), and for forest land, poorly productive forest land and unproductive land, and divided into South and North Finland. Note that on mineral soils forestry land with site fertility categories of 1–6 always belonged to forest land in NFI9, while in the case of peatland forestry land these site fertility categories could also belong to poorly productive forest land and unproductive land.

Over three quarters of the mineral soil forest land in Southern Finland belong to Mesic forests or more fertile categories, and 4% of forest land belonged to the most fertile category herb rich sites (Table A.5). The two most fertile classes are much rarer in North Finland. However, mesic forests are slightly more common in North Finland than in South Finland. The proportion of the combined herb rich forest and herb rich heath forests of mineral soil forest land was highest in South Finland, in the region Häme-Uusimaa, and in some areas of South-East Finland, and in South and North Karelia (Fig. 4.3). The proportion of sub-xeric forests is almost 40% of mineral soil forest land in North Finland and 20% in the south. The proportion of sub-xeric and poorer sites of forest land increased northwards and exceeded 60% in the northernmost part of the country (Fig. 4.4).

4.3 Soil Classification and the Areas of Site Fertility Classes on Mineral Soils

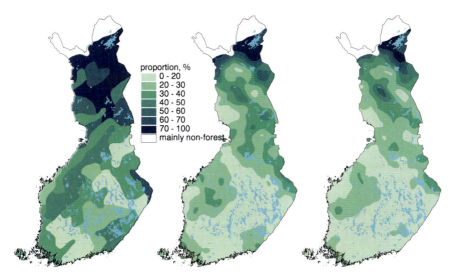

Fig. 4.4 The proportion of sub-xeric heath and less fertile forests of the area of the forest land in the 3rd, 8th and 9th National Forest Inventories

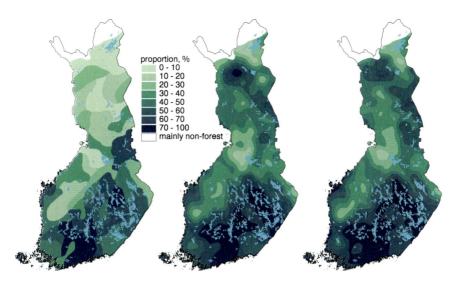

Fig. 4.5 The proportion of mesic heath forests and more fertile forests of the area of the forest land in the 3rd and 8th and 9th National Forest Inventories

It seems that the sites have become more fertile since the beginning of the 1950s, although not necessarily during the period between NFI8 and NFI9. Particularly in North Finland, Mesic forests have become more common than in the early 1950s (see also Figs. 4.3 and 4.5). Some speculations of the reasons are given in Tomppo (2000). Some of the changes may be real and some may be caused by differences in classification methods. The reasons for real changes could be, e.g., regeneration of forests with soil preparations that increase the amount of nitrogen available for plants. A part of the changes could be explained by the slash-and-burn agriculture that was carried out until the end of the nineteenth century. After this type of agriculture, and a short grass and hay phase, the site was poor until the humus layer with nitrogen accumulated. The effect of the accumulated humus layer can be seen now which might explain the changes in South Finland. On the other hand, earlier the former grazing of cattle in forests might also have increased site fertility. The current nitrogen deposition could also have a fertilization effect. In North Finland, reindeer husbandry has decreased the amount of lichen and so the poor sites might seem more fertile than they actually are. The regeneration of forests resulting in large areas of young forests, together with soil preparations, could increase the amount of plants growing on fertile sites. Finally, almost 700,000 ha of drained peatland forests are currently classified as mineral forests. These former peatland forests are often fertile sites and increase the proportion of the fertile site classes.

4.4 Peatlands and Their Site Classes

4.4.1 Peatland Area and Its Changes

A forestry land is classified as peatland if the organic layer on the mineral soil is peat or >75% of the ground vegetation is peatland vegetation (Chap. 2). More than one third of Finland's current land area was peatland before human induced land-use changes. It was still about of one third at the beginning of the 1950s. Some of the earlier peatlands are now in other land-use classes such as agriculture and built-up areas. In NF9, the area of forestry land classified as peatland was 9.062 million ha from which 3.219 million ha were in South Finland and 5.843 million ha in North Finland (Table 4.4, Tables A.5 and A.8). The area of peatland soils has varied to some extent since the beginning of the 1950s, and significantly more than the standard errors of the estimates of the differences. It was 9.742 million ha in NFI3 (1951–1953) and about the same (9.664 million ha) in NFI5 (1964–1970) without the peatlands on the protected areas. The area in NFI8 was 8,912 million ha. The apparent increase in NFI9 was caused partly by the afforestation of arable land and also the fact that soil types were assessed more thoroughly in NFI9 than in NFI8. The peatland soils with a thin peat layer or with vegetation composition typical for upland soils were better indentified as peatland in NFI9 than in NFI8. There are also likely some other differences in soil classifications between inventories causing some variation in the area estimates. The proportion of peatlands of the land area is highest in Central and North Central part of West Finland (Fig. 4.6). It is also high in some regions in East Finland.

The reasons for the decrease of the area of peatland between NFI5 and NFI8 (1986–1994) are analysed and discussed in Tomppo (1999) and Hökkä et al. (2002)

4.4 Peatlands and Their Site Classes

Table 4.4 The areas (1,000 ha) of peatlands by drainage situation and the area of drained mineral soils in the whole country in the 3rd and 5th to 9th National Forest Inventories

Inventory	Field measurement	Undrained	Recently drained	Transforming	Transformed	Total peatland	Drained mineral soils	Total peatland and drained mineral soils
	1,000 ha							
NFI3	1951–1953	8,827	271	507	137	9,742	0	9,742
NFI5[a]	1960–1963	6,607	1,808	884	364	9,664	0	9,664
NFI6	1964–1970	5,273	1,829	1,717	545	9,364		9,364
NFI7	1971–1976	4,546	122	2,582	664	9,012	604	9,626
NFI8	1977–1984	4,231	870	2,902	910	8,912	1,008	9,919
NFI9	1996–2003	4,142	269	3,101	1,550	9,062	1,330	10,392

[a] The areas do not include the peatland on protected areas

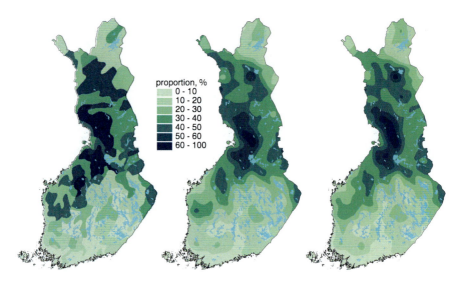

Fig. 4.6 The proportion of the spruce mires, pine mires, and open bogs and fens (peatlands) of the land area in the 3rd and 8th and 9th National Forest Inventories

and further between NFI5 and NFI9 in Tomppo (2005). The most significant reason is that a part of the drained peatlands with a thin peat layer is classified as mineral soil after the peat layer has disappeared following drainage. Other, minor, reasons are the change of land class, due to, e.g., the use of forestry land for roads, other built-up land, water reservoirs or arable land. Reforestation or afforestation of arable land, particularly on organic soils with peat layer, has partly compensated for that decrease. The area transferred from peatland forestry soils to mineral forestry soils between NFI5 and NFI8 due to drainage was assessed to be 660,000 ha by Tomppo (1999) and Hökkä et al. (2002), and between NFI5 and NFI9 510,000 ha by Tomppo (2005). The later figure is lower due to more thorough assessment of soils in NFI9 than in NFI8. There are many factors that make the accurate assessment of the changes difficult, e.g., the incomplete information of land class changes. Permanent field sample plots in earlier inventories had decreased the errors of the change estimates. The transfer has occurred mainly in South Finland where large areas of fertile peatlands with thin peat layer have existed and where the drainage operations began earlier than in North Finland.

4.4.2 Land Classes of Peatlands

Peatland sites are classified in many ways by Finnish forestry and the NFI. The classification bases are, in addition to land-use classes, the principal site class, site fertility class and its specification, as well as the drainage situation (Chap. 2, see also Tables A.5 and A.8).

4.4 Peatlands and Their Site Classes

A part of the peatland drainage operations has taken place on poorly productive and unproductive forestry lands, the latter ones being almost treeless sites. Drainage has lowered the water table and increased the timber productivity and converted sites to forest land. The increase of the forest land area between NFI5 and NFI8, caused by peatland drainage was assessed to be about 1.5 million ha by Hökkä et al. (2002). The conversion has continued between NFI8 and NFI9. The area of forest land on peatland soils was 350,000 ha higher in NFI9 than in NFI8. If it is assumed that about 150,000 ha has reverted from mineral soil forests to peatland forests since NFI8, increasing forest land area on peatlands, it could be concluded that peatland drainage has increased the forest land area between NFI5 and NFI9 by about 1.7 million ha.

The peatland area suitable for drainage and wood production was assessed in earlier forestry programs and forest improvement programs in 1960s to be as high as 7.0–7.5 million ha, e.g., (Kuusela 1972). The drainage of 6.5 million ha was assumed to increase forest area by almost 4 million ha. The ambitions goals set for peatland drainage (to increase the volume of growing stock and its annual increment) were even exceeded even though the drained area was less than set by the highest goals, see Chaps. 7 and 8.

On the basis of NFI9, the area of peatland classified as forest land was 5.15 million ha, 2.67 million ha in South Finland and 2.48 million ha in North Finland (Tables A.5 and A.8). The peatland forest area in NFI5 was 3.93 million ha. As explained earlier, the increase in forest area caused by peatland drainage is higher (1.7 million ha) than the differences of forest areas in NFI9 and NFI5 (1.2 million ha) because drained peatlands with a shallow peat layer convert to mineral soils, and a part of earlier peatland was classified into mineral soils in NFI9. The combined area of poorly productive forest land and unproductive land has decreased at the same time from 5.73 to 3.91 million ha.

4.4.3 Drainage Situation of Peatlands

The drainage situation categories for peatlands in NFI are undrained, recently drained, transforming stage, and peatland forest stage, also called the transformed stage. The peatland forest stage is the final aim of the drainage operation. It is a drained peatland site the ground vegetation of which resembles one of the mineral forests site types and the hydrology does not limit the closure of the stand (Table 2.12). The drainage situation has been assessed in different inventories using the same criteria. The only exception is that the transforming stage had earlier to meet the productivity of forest land. In NFI9, poorly productive forest land or unproductive land could also be classified as peatland forest stage.

In NFI9, the area of drained peatland was 4.92 million ha, and the entire peatland area 9.06 million ha (Table 4.4, Table A.8). When taking into account the area converted from peatland soils to mineral soils due to drainage, the total area of drained peatland soils is about 5.5 million ha. This is close to the area, 5.4 million ha,

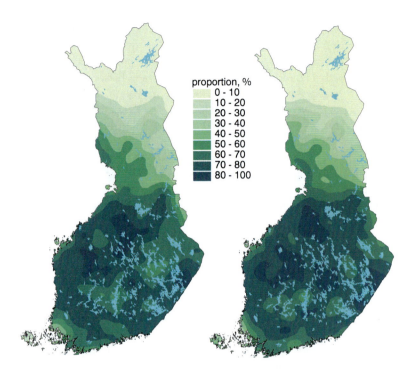

Fig. 4.7 The proportion of drained spruce mires, pine mires, and open bogs and fens (peatlands) of the peatland area in the 8th and 9th National Forest Inventories

based on drainage statistics (Finnish Statistical Yearbook of Forestry 2006). The area of peatland soils in NFI3 and NFI5 was almost 10 million ha (Table 4.4). The proportion of drained peatland of the peatland soils is higher in southern Finland than in northern part of the country and decreases towards North Finland (Fig. 4.7).

In NFI9, a majority (63%) of the drained peatland soils were in a transitional stage and only a minor part were at the stage where the drainage effect was not yet visible (5.5%). The rest (31.5%) were at the peatland forest stage (called transformed in Table A.8). In the southern part of the country, the proportions were 5.7%, 52.5% and 41.8% respectively and in the northern part 5.2%, 73.9% and 20.9%.

The results revealed differences between South and North Finland concerning the proportions of the final peatland forest stage, and also in the rate of conversion of the drained peatland to the final stage. Climatic reasons, the differences in site fertility distributions and, on average, a later time point of the initial drainage in North Finland are the main reasons for the lower share of the peatland forest stage and higher share of transitional stage in North Finland. The proportions of the peatland forest stages had increased between NFI8 and NFI9 in the entire country but were still low. The reasons could be, in addition to conversion speed, the drainage of peatlands of poor site fertility and also relatively sparse ditch spacing in some regions (cf. Hökkä et al. 2002).

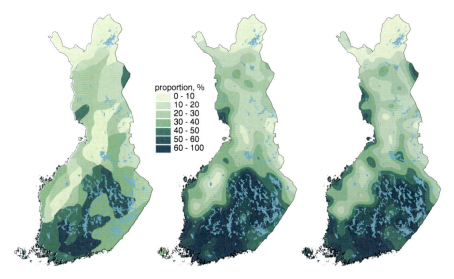

Fig. 4.8 The proportion of spruce mires of the peatland area in the 3rd, 8th and 9th National Forest Inventories

4.4.4 Principal Site Classes and Site Fertility Classes on Peatland Soils

The main site classes are: (1) Forest in mineral soil, (2) Spruce mires, (3) Pine mires, (4) Open bog or fen (Treeless peatland). Drained peatlands that have been treeless peatlands in their natural states, but forested naturally or through artificial regeneration, are also classified as spruce or pine dominated peatlands. Note that scattered trees can occur on open bogs and fens.

The distribution of the principal site classes and site fertility classes is examined separately for undrained and drained peatland. The drainage operations change treeless peatland to either spruce mires or pine mires, depending on the site. The site fertility class, however, remains unchanged when the site is transforming from undrained stage to peatland forest stage.

The current undrained peatland soils are likely to remain so, and their area (4.1 million ha) will remain constant. The first-time drainage of undrained peatland soils has practically ended in Finland. The Finnish Forest Certification Systems also forbids the drainage of pristine peatland soils (FFCS 2006). In NFI9, the area of undrained spruce mires was 0.8 million ha, pine mires 1.8 million ha and open bogs or fens 1.6 million ha (Figs. 4.8–4.10). The areas and the proportions of peatland areas of undrained pine mires and open bogs and fens were highest in North Finland (cf. also Fig. 4.9). The distribution of undrained peatland soils into principal site classes has changed since the beginning of 1950s (NFI3, 1951–1953) following drainage operations, in such a way that the proportions of spruce and pine mires have decreased and that of open bogs and fens increased (cf. Ilvessalo 1956).

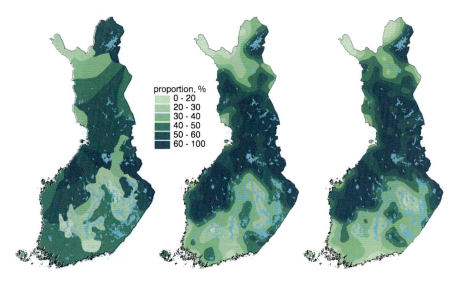

Fig. 4.9 The proportion of pine mires of the peatland area in the 3rd, 8th and 9th National Forest Inventories

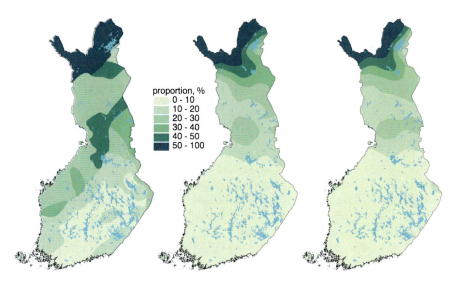

Fig. 4.10 The proportion of open bogs and fens of the peatland area in the 3rd, 8th and 9th National Forest Inventories

The distributions of peatland soils by site fertility classes are shown in Table A.5. The differences in the distributions of site fertility classes of the undrained peatland soils in South and North Finland reveal the differences in natural conditions and also the differences in selecting peatland soils of different site fertility classes for drainage.

4.4 Peatlands and Their Site Classes 111

In South Finland, the proportion of spruce mires of the all drained mires was significantly higher than in North Finland. About 75% of all drained peatland soils in North Finland were pine mires. The site fertility distribution of the drained peatlands deviated surprisingly little from that of undrained peatland soils. Some regional differences were observed in the prevalence of drained spruce and pine mires with site fertility of middle classes in South Finland while in North Finland, the typical drainage objects have been fertile sites, e.g., herb-rich spruce mires and tall sedge pine mires. However, it should be noted that the principal site class distribution and site fertility class distribution of the drained mires at the inventory time point do not reveal the original distribution of the principal site classes and their site fertility distributions due to the fact that open fens were classified as either spruce or pine mires once the forest growing stock had appeared. The proportion of spruce mires of all drained peatlands at the time point of NFI9 was 33% (1.6 million ha), pine mires 66% (3.3 million ha) and open bogs or fens 1% (60,000 ha). The respective proportions of the undrained peatlands were 18% (770,000 ha), 44% (1.8 million ha) and 38% (1.6 million ha) (Tomppo 2005).

4.4.5 The Thickness of the Peat Layer

In the NFI9, the thickness of the peat layer was measured to a maximum thickness of 4 m (Sect. 2.2.2). Such a thickness was used before NFI9 in NFI3 (1951–1953) (Ilvessalo 1956). In NFI8 (1986–1994) in South Finland, it was only recognised whether the thickness of the organic layer was less or more than 30 cm. In North Finland in NFI9, the thickness was measured up to 100 cm. The change was made in the connection with other changes in NFI8. In NFI9, the changes were a result of the need to be able to estimate carbon stocks accumulated in soils, as well as carbon stock changes. Furthermore, the data can be used in growth studies, as well as in assessing peat reservoirs. The peat layer has accumulated since the last ice age, in about little less than 10,000 years. Peatland drainage has changed the thickness of the peat layer to some extent since the beginning of 1950s. The thickness of peat layer varies significantly within the country, mainly from South to North but also from West to East (Ilvessalo 1956, Table A.7).

Classified on the basis of principal site class, spruce mires had the thinnest peat layer. In the entire country, 45% of the area of the spruce mires had a peat layer not thicker than 30 cm (Table A.7). The thickness distributions were similar in South and North Finland. Spruce mires were often rich of nutrients. The nutrient balance, and particularly the level of water table, may prevent or reduce wood production without drainage operations. Spruce mires were often classified as forest land. The average thickness (when the measurement was terminated at 4 m) was 65 cm. Note that only 1.3% of the area of spruce mires had a peat layer thicker than 4 m wherefore the average of 65 cm is a nearly unbiased estimate.

Table 4.5 Peatland areas (1,000 ha) by the thickness class of peat layer in the whole country in in the 3rd and 9th National Forest Inventories

Thickness of peat (cm)	NFI3[a] Area 1,000 ha	%	NFI9 Area 1,000 ha	%
–30	2,181	22.4	2,252	24.9
30.1–50	1,577	16.2	1,396	15.4
50.1–100	2,094	21.5	2,002	22.1
101–200	2,465	25.3	1,922	21.2
201–300	880	9.0	826	9.1
301–400	371	3.8	301	3.3
401–	175	1.8	364	4.0
Total	9,742	100.0	9,062	100.0

[a] The areas excludes drained peatlands (9.4% of the peatland area)

The thickness of the peat layer of pine mires was, on average, significantly higher than that of spruce mires, 21% of the area of pine mires had a thickness of not more than 30 cm, and 17% thicker than 2 m. The peat layer of pine mires was on the average noticeably thicker in South Finland than in North Finland. The average thickness, when the measurement was terminated at 4 m, was 123 cm in the entire country, 150 cm in South and 98 cm in North Finland. In the South Finland, 7.5% of the area of pine mires had a peat layer thicker than 4 m, so the real average thickness was somewhat greater than 150 cm. This was true for the whole country (3.7% of the mires with a higher peat depth than 4 m) and to some extent to North Finland (1.7% thicker than 4 m).

The average thickness of the peat layer of open bogs and fens was even greater than that of pine mires. Almost one third of those sites in South Finland had a peat layer thicker than 4 m. The average thickness in the entire country was >193 cm, >252 cm in South and >172 cm in North.

When comparing the thickness of the peat layer measured by NFI3 and NFI9, it should be remembered that during the interim a significant part of the peatlands were converted to other land-use classes, and particularly to mineral soil forests due to a thin peat layer. Further, large areas of open fens were drained and therefore converted to spruce or pine mires. The exact area of the drained open fens or bogs is somewhat difficult to asses due to the lack of permanent plots. However, the area of the open fens or bogs in NFI3 (1951–1953) was 2.72 million ha while in NFI9 it was 1.62 million ha (Ilvessalo 1956, see also Hökkä et al. 2002, and Tomppo 2006). There are also other reasons for the diminished area of open fens and bogs than the drainage of those sites, e.g., the lowering of the water table due to the drainage in the surrounding areas. Note that the area of the drained mires was 920,000 ha already in NFI3. The distribution of peatlands into spruce mires, pine mires and open bogs and fens in NFI9 was different from that in NFI3. Nevertheless, the thickness of the peat layer by main site classes in NFI3 and NFI9 were quite close to each other, particularly for spruce mires (Table 4.5, Ilvessalo 1956).

4.5 Tree Species Dominance and Composition

4.5.1 The Dominant Tree Species

The principles for assessing the proportions of tree species and the dominant tree species of a stand are described in Sect. 2.11. As well as the peatland drainage, one of the biggest changes in the Finnish forests since the beginning of 1950s has been the change in tree species dominance. The area of pine dominated forests has increased significantly. The main reasons are the regeneration of spruce dominated forests with pine and favouring of pine in planting or seeding. To minor extent the establishment of new forests through peatland drainage has also favoured pine. The favouring of pine was intentional throughout the 1960s and 1970s because of the assumption that high quality pine saw timber would have export markets in the future. When examining the changes in tree species dominance, it should also be kept in mind that the forest land area has increased since the beginning of 1950s and that the current national definition of forest land has been in use only since NFI5 (1964–1970). The current forest definition includes poorer sites than the definition in NFI3 (1951–1953). These poorer sites are often either pine mires or characterised by pine dominated poor mineral soils.

According to NFI9, the area of Scots pine (*Pinus sylvestris* L.) dominated forests in the entire country was 13.3 million ha. It was 9.2 million ha in NFI3 and in 10.4 million ha in NFI5 (Table 4.6, Fig. 4.11, Table A.9). The area in South Finland has increased since NFI5 (1964–1968) from 5.0 million ha to 6.3 million ha and in North Finland (1969–1970) absolutely and relatively even more, from 5.4 to 7.0 million ha. In the country as a whole, 65% of forest land was pine dominated and in North Finland 76% (see also Fig. 4.12). The proportion of pine dominated forests of the forest land area was higher than the proportion of pine volume from the volume of growing stock (Sect. 4.7). The reason is that pine dominated forests are, on the average, younger that spruce dominated forests both in South and North Finland.

The area of spruce dominated forests (*Picea abies* (L.) Karst.) had decreased since the end of 1960s by 1.2 million ha, proportionally more in North Finland, and is now 4.8 million ha. In South Finland, 31% of forest land area was spruce dominated, while the proportion in North Finland was 15%, see also Fig. 4.13. The rate of decrease is levelling off and it is likely that the area of spruce dominated forests will not continue to decrease as rapidly as in the 1970s and 1980s.

In NFI9, the proportion of birch dominated forests of forest land (Silver birch, *Betula pendula* Roth and Downy birch, *Betula pubescens* Ehrh.) in the entire country was 8.7%, 9.9% in South Finland and 7.4% in North Finland (Tables A.9 and A.10). In South Finland, Silver birch dominated forests were almost as common as those dominated by Downy, while in North Finland, Silver birch was rarely a dominant species. Because of slash-and-burn agriculture, birch has been common in Eastern Finland because of its role as a pioneering species after such an agriculture (Fig. 4.14).

Table 4.6 The area of forest land (1,000 ha) by dominant tree species in the 3rd and 5th to 9th National Forest Inventories for the entire country and in the 5th to 9th National Forest Inventories for South and North Finland

Inventory	Field measurement years		Temporarily unstocked area	Pine	Spruce	Birch	Other broadleaved trees	Total
Whole country								
NFI3	1951–1953	1,000 ha	260	9,197	4,963	2,568	364	17,352
		%	1.5	53.0	28.6	14.8	2.1	100.0
NFI5	1964–1970	1,000 ha	804	10,377	5,983	1,346	187	18,697
		%	4.3	55.5	32.0	7.2	1.0	100.0
NFI6	1971–1976	1,000 ha	790	11,488	5,941	1,283	237	19,738
		%	4.0	58.2	30.1	6.5	1.2	100.0
NFI7	1977–1984	1,000 ha	629	12,417	5,445	1,407	169	20,067
		%	3.1	61.9	27.1	7.0	0.8	100.0
NFI8	1986–1994	1,000 ha	297	12,991	5,146	1,504	135	20,074
		%	1.5	64.7	25.6	7.5	0.7	100.0
NFI9	1996–2003	1,000 ha	263	13,319	4,829	1,779	147	20,338
		%	1.3	65.5	23.7	8.7	0.7	100.0
South Finland								
NFI5	1964–1968	1,000 ha	409	4,998	4,189	826	169	10,591
		%	3.9	47.2	39.6	7.8	1.6	100.0
NFI6	1971–1974	1,000 ha	399	5,275	430	737	215	10,930
		%	3.6	48.3	3.9	6.7	2.0	64.6
NFI7	1977–1982	1,000 ha	286	5,974	3,943	725	154	11,071
		%	2.6	54.0	35.6	6.5	1.4	100.1
NFI8	1986–1992	1,000 ha	176	6,259	3,747	793	116	11,091
		%	1.6	56.4	33.8	7.1	1.0	100.0
NFI9	1996–2000	1,000 ha	163	6,332	3,442	1,102	129	11,167
		%	1.5	56.7	30.8	9.9	1.2	100.0

4.5 Tree Species Dominance and Composition

North Finland											
NFI5	1969–1970	1,000 ha	396		5,379		1,794		520	18	8,106
		%	4.9		66.4		22.1		6.4	0.2	100.0
NFI6	1975–1976	1,000 ha	391		6,213		1,638		546	22	8,808
		%	4.4		70.5		18.6		6.2	0.2	100.0
NFI7	1982–1984	1,000 ha	343		6,444		1,511		682	15	8,996
		%	3.8		71.6		16.8		7.6	0.2	100.0
NFI8	1992–1994	1,000 ha	121		6,733		1,398		712	19	8,983
		%	1.3		75.1		15.5		7.8	0.3	100.0
NFI9	2001–2003	1,000 ha	100		6,988		1,387		678	18	9,171
		%	1.1		76.2		15.1		7.4	0.2	100.0

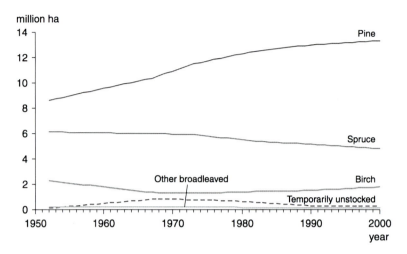

Fig. 4.11 The tree species dominance on forest land in the 3rd to 9th National Forest Inventories

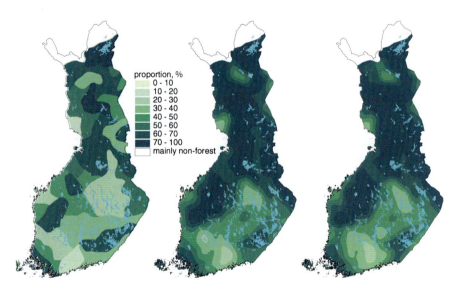

Fig. 4.12 The proportion of pine (*Pinus sylvestris* L.) dominated forests of the forest land area in the 3rd, 8th and 9th National Forest Inventories

The other broadleaved trees are mainly aspen (*Populus tremula* L.), grey alder (*Alnus incana* (L.) Moench) and common alder (*Alnus glutinosa* (L.) Gaertner). In NFI9, the areas of forest land dominated by these species were small. The other tree species occurred as dominant species only occasionally, and then mostly in the southernmost part of the country, particularly in Ahvenanmaa (see also Table A.9).

The relative decreases in the areas of birch dominated forests and forests dominated by other broadleaved trees since the NFI5 (1964–1970) were even higher

4.5 Tree Species Dominance and Composition

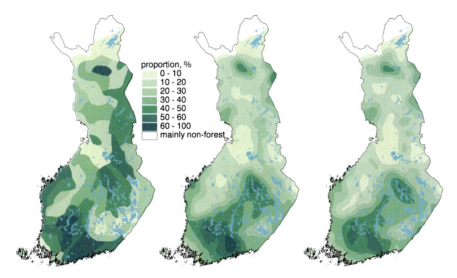

Fig. 4.13 The proportion of spruce (*Picea abies* (L.), Karst.) dominated forests of the forest land area in the 3rd, 8th and 9th National Forest Inventories

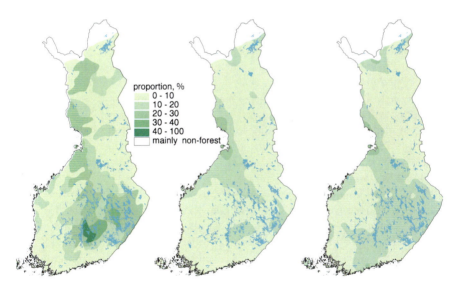

Fig. 4.14 The proportion of birch (*Betula* spp.) dominated forests of the forest land area in the 3rd, 8th and 9th National Forest Inventories

than that of spruce dominated forests. However, the volumes had increased between NFI3 and NFI9, although the volume of birch first decreased during the 1960s and 1970s. The broadleaved trees therefore occurred more often as a mixture in pine and spruce dominated forests than in the 1950s and early 1960s.

In NFI9, the area of temporarily unstocked forests, i.e., forests under regeneration, was 263,000 ha, 1.3% of forest land area. It was highest, 4% of forest area, during the intensive regeneration phase in late 1960s and in 1970s. The proportion of unstocked forests in NFI9 was higher in South Finland (1.5%) than in North Finland (1.1%). The reasons could be a shorter rotation in South than in North Finland, and the proportion of protected forest of forest area is much higher in North than in South Finland.

4.5.2 Tree Species Dominance by Site Fertility Classes

In Boreal forests in natural state, the tree species dominance and tree species composition are largely determined by the site fertility and climate. The forestry regimes since the 1950s have changed the dominance to some extent. In natural forests, Herb rich and Herb rich heath forests are often dominated by spruce or broadleaved trees, and Mesic forests are often spruce dominated. In South Finland, 18% of Herb rich and Herb rich heath forests, and 56% of Mesic forests were pine dominated in NFI9. The proportions in North Finland were 23% and 67%, respectively. Similar shares in NFI2, assuming contemporary boundaries between South and North Finland, the proportions were 16% and 31% in South Finland and 3% and 21% in North Finland. There has been s strong tendency from spruce dominated to pine dominated forests, particularly in North Finland.

At the beginning of the 1950s, pine dominated forests were still common in those regions in which Sub-xeric forests and poorer sites were also common (Fig. 4.15, see also Figs. 4.4 and 4.12). However, a part of Mesic forests had been regenerated with spruce at that time. For example, 40%, in eastern part of South Finland, while in many areas of the country 18% of Mesic forests were pine dominated (Ilvessalo 1956, 1960). Pine dominated forests have become more common since then. The increase was high in South-East Finland, in the Pohjanmaa region, and in the entire North Finland.

At the beginning of 1950s, the proportion of spruce dominated forests of the area of the productive forest land was highest in South and South-East Finland, the coastal region of the Gulf of Bothnia, in some areas of northern and central South Finland, as well as in northern central Lapland (Figs. 4.13 and 4.15). The sites favoured by spruce, Mesic and more fertile sites, were common in those areas. In NFI9, the proportion of spruce dominated forests was highest in the southern part of the country. The proportion had increased between NFI3 and NFI8 in some areas in which birch dominated forests were earlier common, e.g., east of lake Päijänne.

Poorly productive forest lands were typified by pine mires, or till mineral soils in South Finland and rocky mineral soils in North Finland (Tables A.5 and A.6). Poorly productive forests were common in North Finland and so the tree species distribution on poorly productive forests in the entire country followed that of North Finland. Pine was the most common dominant tree species on poorly productive

4.5 Tree Species Dominance and Composition

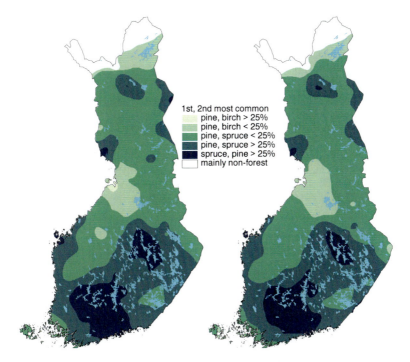

Fig. 4.15 The two most common dominating tree species on forest land in the 3rd, 8th and 9th National Forest Inventories, and the proportion of forests dominated by the second most common dominating species

forest land in both South and North Finland. The second commonest tree species on those sites was Downy birch. It occurred in South Finland mostly on mires and in North Finland also on mineral soils.

4.5.3 Tree Species Mixtures

Forest stands were divided into three classes according to the composition of tree species: (1) the proportion of dominant tree species (coniferous or broadleaved) was over 95% (pure or almost pure single species stand), (2) the proportion of dominant (coniferous or broadleaved) species was equal to or more than 75% but not more than 95%, and (3) the proportion of dominant (coniferous or broadleaved) tree species was less than 75% (Table A.10).

Of the pine dominated forests, somewhat less than half (46%) in South Finland and somewhat more than half (53%) in North Finland were pure or almost pure single species stands (more than 95% pine) (Table A.10). The favouring of pine regeneration in Mesic forests had increased the area of pine forests with mixtures.

Pure or almost pure single species stands are much rarer among forests dominated by other tree species than pine.

The share of such forests in which the proportion of a single species is less than 75% was 30% in South and 25% in North Finland. The broadleaved dominated forests, both birch and other broadleaved trees, were such that the proportion of non-dominant species was generally more than 25%. These types of forests were rarest among pine dominated forests (Table A.10).

On the basis of international definitions, mixed forest or other wooded land is such that neither coniferous nor broadleaved trees account more than 75% of the tree crown area (UNECE/FAO 2000). A modified version of this definition can be used in such a way that the tree species proportions are determined as in the Finnish NFI, i.e., based on the proportions of volume in young thinning stands and more mature stands, and based on stem number capable for development on seedling stands (Chap. 2 and Sect. 4.5.1). Using this modified definition, the share of mixed forests in NFI9 was 14% in South Finland and 13% in North Finland. Mixed forests were rare among pine dominated forests (10% in South and 8% in North Finland). Mixed forests were relatively common among birch dominated forests, and in North Finland also among forests dominated by other broadleaved tree species.

The tree species mixture was assessed in NFI9 in terms of the second most important tree species. Its proportion was assessed in the field with the reference to the proportion of the dominant tree species. Note that the most important second species was not necessarily the second commonest species on seedling stands where no tending was carried out. The second species was recorded in South Finland for 55% of the area of pine dominated forests, i.e., an area of 3.45 million ha from 6.93 million ha. The commonest second species in pine dominated forests in South Finland was spruce which occurred on 25% of the area of pine dominated forests (Table A.11). Its proportion was in most cases less than 25%. The second commonest species was downy birch. In spruce dominated and broadleaved dominated forests, a second species was more common than in pine dominated forests. The commonest most important second species in spruce dominated forests was pine and the second commonest species downy birch.

In North Finland, a second species was relatively rarer than in South Finland in pine dominated forests, and more common in spruce dominated forests, being about as common as in South Finland in birch dominated forests (Table A.11). The commonest most important second species in pine dominated forests was spruce and the second commonest downy birch, as in South Finland. In spruce dominated forests, downy birch and pine were almost as common in about 40% of the area of the spruce dominated forests. In birch dominated forests, the commonest second species was pine and second commonest spruce.

The changes in the abundance of the commonest and second commonest tree species were rather small between NFI8 (1986–1994) and NFI9 (1996–2003). The area in which spruce is the commonest species had decreased slightly (Fig. 4.15). The other minor change is that birch had become slightly more common as the second species in pine dominated forest between the two inventories.

4.6 Age and Development Classes

4.6.1 The Age Distributions of Stands and Their Changes

The age of a stand means the age of the trees of the stand. An assumption behind the concept of age is that the trees in the stand are of the same age. In the boreal forests, even in artificially regenerated stands, trees with different ages exist due to natural germination. Thus, in the boreal forests, the age has to be assessed as a weighted average of individual trees, see Sect. 2.11.

Optimized wood production presumes an even age class distribution in a forest area, with the oldest stands being as old as the rotation. However, different sites with different fertilities may require different species and somewhat different rotation times. Other functions of the forest may also result in an uneven age distribution. More importantly, different forest owners, in practice treat their forests in different ways and for different purposes. Thus, an even age distribution is seldom attained.

In the first half of the 1900s, the forests in many areas were subjected to selective cuttings. In South Finland, the forest stand age classes 41–60 and 61–80 years were more common than in the equal area distribution would had been still at the beginning of the 1950s, and the 61–80 class still at the beginning of 1960s (Tomppo and Henttonen 1996). The age distribution has become more even after that (Fig. 4.16). The proportions of age classes 1–20 and 21–40 years had increased since NFI5 (see also Fig. 4.17). The proportions of the stands with the age of 61–80 and 81–100 years had decreased at the same time, while the proportion of forests older than 100 years had been continuously slightly increasing (see also Fig. 4.18). These trends are a consequence of the changing forestry practices, i.e. the shift from selective cuttings to rotation-based forestry. The reluctance of some forest owners to sell timber is one reason for the increase of the area of older forests, in addition to a slight increase in protected forests in South Finland.

The proportion of temporarily unstocked regeneration areas had decreased steadily in South Finland since the NFI5 (1964–1968) when it was 4.0% of the area of the forest land. It was 1.5% in NFI9. The establishment of new forest has therefore become faster since 1970s.

When looking the age class distributions by tree species or tree species groups the age class distribution is no longer even. The area of the 21–40 years age class was the highest among the pine dominant forests in South Finland, i.e., the class is the mode. The area was significantly larger than the average in forests in South Finland (Table A.12a). The regeneration of earlier spruce dominated forests with pine since the 1960s is one factor affecting the age distributions by tree species in NFI9. The mode age class of spruce dominated forests was 61–80 years. The area of the age class 81–100 years was also more common in spruce forests than the areas of other age classes. Broadleaved dominated forests were significantly younger than coniferous forests. A shorter rotation of broadleaved trees is one explanation for the younger age classes of broadleaved tree species dominated stands compared to coniferous stands. Other reason is the long lasting tradition to

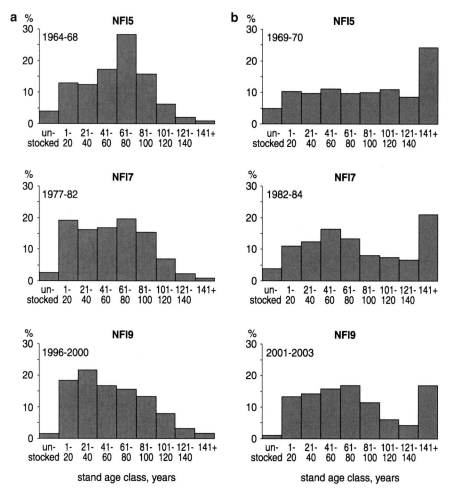

Fig. 4.16 The age class distribution on forest land (**a**) in South Finland and (**b**) in North Finland in the 5th, 7th and 9th National Forest Inventories

use coniferous species, particularly pine, as regeneration material. In recent years, the tradition has changed to some extent and the use of spruce and broadleaved trees has increased.

The mean volume (m³/ha) and mean diameter of trees increased with the increased age until the 81–120 years age class in South Finland. After that, the mean volume did not increase while mean diameter increased slightly or remained as unchanged (Tables A.12b and c). Note that these parameters were estimated from tally trees by stand age classes, not, e.g., using the assessed mean diameter of stand (Sect. 3.2.1 and 3.2.2). Another interesting finding was that the mean volume was higher in spruce dominated than in pine dominated forests in age classes 21–40 years or older. This was also the case for the mean diameter until the age class 101–120 years.

4.6 Age and Development Classes

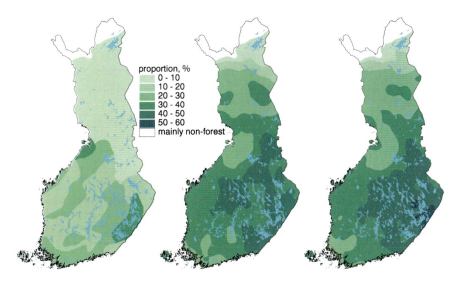

Fig. 4.17 The proportion of less than 40 years old forests of the forest land area in the 3rd, 8th and 9th National Forest Inventories

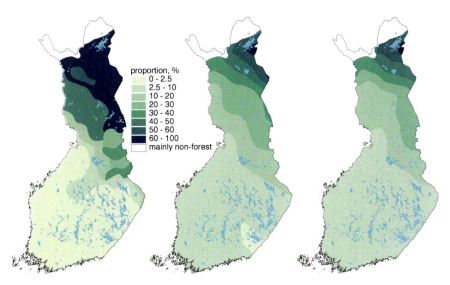

Fig. 4.18 The proportion of over 120 years old forests of the forest land area in the 3rd, 8th and 9th National Forest Inventories

The age distribution of forest stands in the 1920s, and still at the beginning of 1950s, was very different in North Finland from that in South Finland. At the beginning of 1920s, over one half of the forests were older than 140 years (Ilvessalo 1927). The reason was the less intensive use of timber resources over wide areas of the

North Finland, and the fact that the practice of regeneration cuttings reached the northern part of the country later than the southern part. Still in NFI5 (1969–1970), one third of the forests were older than 120 years. In NFI7 (1982–1984), it was about 30%, and 21% in NFI9 (2001–2003). The old forests were common in state owned forests where the wide areas of protected forests increased their proportion. The proportion of young forests (1–40 years old) has increased continuously since the late 1960s, but has probably reached its maximum in NFI9 (Fig. 4.16). The proportion of forests under 60 years of age was 44% (Table A.12a). The proportion of forests over 140 years old was 17%. Old forests are still common in East and North-East Lapland (Fig. 4.18).

The pine dominated forests are even younger in North Finland than in South Finland due to the extensive regeneration of old and often sparsely stocked spruce dominant forest since 1950s. The share of forests with an age less than or equal to 80 years is 64% of the area of pine dominated forests (Table A.12a). Spruce dominated forests are older than pine dominated forests. The proportion of spruce dominated forests over 160 years old was 37%. A significant proportion of these forests was protected; the corresponding proportion in forests available for wood production was 23%.

The area of the temporarily unstocked regeneration stands increased rapidly in 1950s and 1960s due to the intensified forestry caused by the new forestry programs. These regimes resulted in extensive forest regeneration in North Finland. This was possible by to the large areas of old forests. The proportion of the unstocked regeneration stands of forests land was 4.9% in NFI5. It has decreased steadily and was 1.1% in NFI9.

As in South Finland, in North Finland, the mean volume of trees (m^3/ha) and mean diameter of trees increased with age until the 141–160 years age class. The mean diameter was highest in class over 160 years (Tables A.12b and c). Spruce dominated forests also had higher mean volumes than pine dominated forests in North Finland, and also by age classes in the age classes 41 years and older. Forests dominated by young broadleaved trees are very dense and have a high mean volume, but not a higher mean diameter, than coniferous forests.

4.6.2 The Development Classes of Stands and Their Changes

Development class of stands have been assessed in NFIs since NFI3 and published since NFI4. The development class describes the development phase of a stand in rotation-based forestry. It is a useful parameter when assessing possible cutting and silvicultural treatments of a stand.

Some target distributions of the development classes for the optimal wood production in a large area had been presented for practical forestry. One distribution, often cited, is that the common proportion of open regeneration forests, young and advanced seedling stands, as well as seed tree and shelterwood stands should be 25%, the proportions of both young and advanced thinning stands 30%, and the

4.6 Age and Development Classes

Table 4.7 Proportions of the areas of the development classes on forest land in the 8th and 9th National Forest Inventories

	South Finland		North Finland	
Inventory	NFI8	NFI9	NFI8	NFI9
Field measurement years	1986–1992	1996–2000	1992–1994	2001–2003
Development class	Proportion of forest land (%)			
Temporarily unstocked areas	1.6	1.5	1.3	1.1
Young seedling stands	6.2	6.6	6.9	5.9
Advanced seedling stands	13.3	14.9	16.2	14.5
Young thinning stands	32.3	33.6	38.1	41.5
Advanced thinning stands	28.1	26.9	14.7	17.2
Mature stands	16.7	14.9	20.6	18.1
Shelter tree stands	1.0	0.4	0.8	0.3
Seed tree stands	0.8	1.2	1.4	1.3
Forest land total	100.0	100.0	100.0	100.0
Area of forest land, 1,000 ha	11,091	11,167	8,983	9,171

proportion of mature stands 15%. The distribution of the development class in NFI9 in South Finland was close to the optimal except that the proportion of young thinning stand was higher and advanced thinning stands lower than in the optimal distribution (Table 4.7, Table A.13a). The areas and proportions of mature stands and advanced thinning stands had slightly decreased between NFI8 and NFI9 in South Finland due to increased cutting activities in the second half of the 1990s. The open regeneration area had not increased, which indicates that regeneration after regeneration cutting has been was faster than in the past decades (Table 4.7). The mean volumes had increased in all development classes between NFI8 and NFI9, for example, from 81 to 99 m^3/ha in young thinning stands, from 181 to 190 m^3/ha in advanced thinning stands and from 208 to 234 m^3/ha in mature stands (Tomppo et al. 2001, Table A.13a).

As in case of age classes, the development class distributions in all forests and in forests available for wood production are similar (Tables A.13a and b). Basic forest characteristics, such as basal areas, mean volumes and mean diameters by development classes are also presented for forests available for wood supply (Table A.13b). These characteristics are indicators typically used in managing forests for wood production. Pine dominated forests, and particularly birch and other broadleaved tree dominated forests, were generally concentrated into younger development classes than spruce dominated forests, which could be deduced also from the age class distributions. In South Finland, 40% of pine dominant forests belonged to the young thinning stands development class and 13% to mature stands. This situation is a consequence of wide forest regeneration operations since the 1950s, particularly in the period from 1960 to 1990 as well as favouring of pine in regeneration. Some new peatland forests were also pine dominated. Of spruce dominated forests, 40% belonged to advanced thinning stands, and about one fifth to both young thinning stands and mature stands.

Two major factors affecting the development class distribution of forests in North Finland were the extensive regeneration areas characteristic of past decades as well as extensive areas of protected forests. The result was a high proportion of young thinning stands, and on the other hand a relatively high proportion of mature forests (42% and 18% respectively) (Table 4.7, Table A.13a). The situation was quite different in forests available for wood production. There the proportion of young thinning stands was 45% and that of mature forests 12%. A somewhat small proportion of advanced thinning stands (18%) could be a problem for sustainable long-term wood production for some decades to come. This was expectable because reaching a balanced distribution takes several decades after the extensive regeneration activities of past decades. The mean volume and basal area of trees had increased in young thinning stands between NFI8 and NFI9 (mean volume from 54 to 64 m^3/ha), and less so in mature forests (from 110 to 114 m^3/ha), but remained about in same in advanced thinning stands (Table A.13a; Tomppo et al. 2001).

Pine dominated wood production forests in North Finland were even younger than in the south, with 48% of the area belonging to young thinning stands, while spruce dominated forests were older than in South Finland. Birch dominated and other broadleaved trees dominated forests were young. The area of the latter one was very small in North Finland.

4.7 Growing Stock

4.7.1 Mean Volume Estimates by Tree Species

According to NFI9, the mean volume of growing stock on combined forest land and poorly productive forest land was 91 m^3/ha and on forest land 100 m^3/ha. The sampling error of the mean volume estimates was 0.4 m^3/ha in both cases (Table A.15a). In South Finland, the mean volume on combined forest land and poorly productive forest land was twice of that North Finland (Table A.15a; Fig. 4.19). Almost half of the growing stock was pine. The mean volume of pine decreases from South to North due to climatic reasons and also due to the history of forestry. The mean volume was highest in the south-western coastal region and also in east parts of South Finland where pine dominated forests are common (Fig. 4.20). Pine forests are young in North Finland a fact that, in addition to climatic factors, also decreases the mean volume there. The regional variation of the mean volume of spruce is even higher than that of pine. The highest mean volumes were in south-western Finland and in the northern lake district area (Fig. 4.21). Relatively high mean volumes of spruce also occur in small areas in North Finland and elsewhere in South Finland. The high mean volumes of spruce coincide with geographical areas where rich site classes are abundant. The mean volume of spruce in South Finland is four times that in North Finland (Table A.15a). The highest mean volumes of birch were found in the lake district, while the other broadleaved tree species were concentrated in South Finland (Figs. 4.22 and 4.23).

4.7 Growing Stock

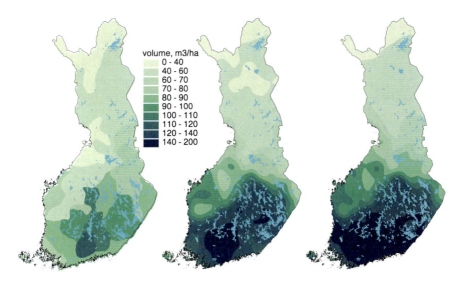

Fig. 4.19 The mean volume of growing stock (m³/ha) on combined forest land and poorly productive forest land in the 3rd, 8th and 9th National Forest Inventories

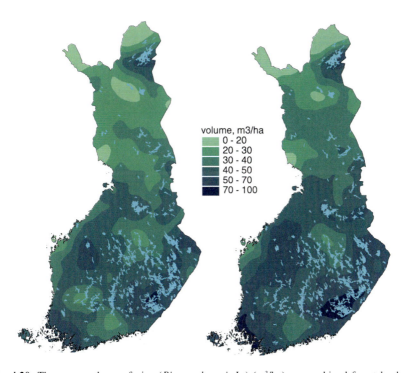

Fig. 4.20 The mean volume of pine (*Pinus sylvestris* L.) (m³/ha) on combined forest land and poorly productive forest land in the 8th and 9th National Forest Inventories

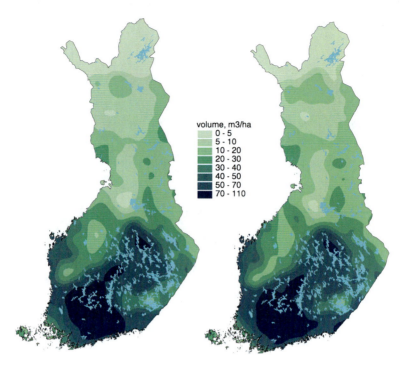

Fig. 4.21 The mean volume of spruce (*Picea abies* (L.), Karst.) (m^3/ha) on combined forest land and poorly productive forest land in the 8th and 9th National Forest Inventories

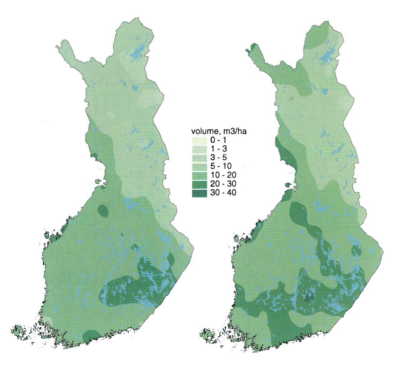

Fig. 4.22 The mean volume of birch (*Betula* spp.) (m^3/ha) on combined forest land and poorly productive forest land in the 8th and 9th National Forest Inventories

4.7 Growing Stock

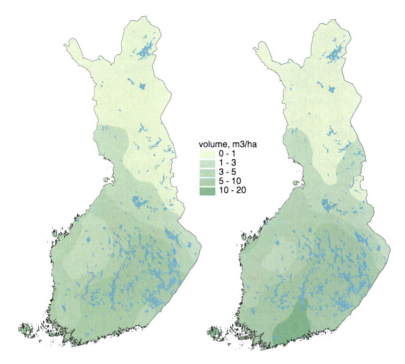

Fig. 4.23 The mean volume of other broadleaved trees than birch (m³/ha) on combined forest land and poorly productive forest land in the 8th and 9th National Forest Inventories

Table 4.8 The mean volume on forest land in the 5th to 9th National Forest Inventories

Tree species	NFI5[a] 1964–1970 m³/ha	NFI6 1971–1976	NFI7 1977–1984	NFI8 1989–1994	NFI9 1996–2003
Pine	33	33	36	42	48
Spruce	29	29	30	34	34
Broadleaved	14	13	15	16	19
Total	77	75	81	92	100

[a] Converted to correspond the new volume models, 3% increase in the volume estimates

The mean volume on forest land (note the different land class to that in Figs. 4.19–4.23) has clearly increased since the end of 1970s (Table 4.8) (Kuusela 1972, 1978; Kuusela and Salminen 1991; Tomppo et al. 2001). It is not possible to make exact comparisons between the mean volumes in 1950s or earlier and those in NFI5 and later because the definition of forest land was changed in NFI5. The extensive regeneration cuttings of old forests in the 1960s and 1970s kept the mean volume low until NFI7 (cf. total growing stock in Sect. 4.7.2.1). However, the increase in mean volume on combined forest land and poorly productive forest land since NFI3 is largest in South Finland (Fig. 4.19).

Note that new volume equations have been employed since NFI6, yielding an approximately 3% increase in the growing stock estimates than the models used in NFI3 and NFI5. The volumes of NFI3 and NFI5 have been converted to correspond to the new volumes. The trees thinner than 2.5 cm at $d_{1.3}$ have been measured only since NFI7, an exception was the South Finland in NFI1. The proportion of small trees of the growing stock in NFI7 was approximately 0.7% (Kuusela and Salminen 1991).

Intensified forestry with nation level forestry programs since the 1960s have resulted in soil improvements and changed silvicultural practices with considerable consequences for Finnish forests (cf. Sects. 1.4, 4.5 and 4.6). The practice of selective cutting that had prevailed for a long period was changed to rotation based forestry. Forests with low density and low increment were regenerated with Scots pine. Uneven-aged forests therefore became even-aged ones. The age structure and tree species composition also changed drastically, particularly in North Finland. As the new dense forests became older, the mean volume of forests also started to increase.

Note that still in the 1950s, the mean volume of growing stock was relatively low due to the intensive use of forest in the nineteenth century and expanding forest industries in the early twentieth century. Most part of the increase has happened on the forests on mineral soils. However, peatland drainage for lowering the water table to improve the growth conditions of trees has also had an impact on the increase of volume increment (see also Sect. 4.8). The growing stock on site classes that were peatland in 1950s (NFI3) had more than doubled in 50 years, by NFI9 (Tomppo 2006).

The favouring of pine in artificial regeneration had increased the mean volume of pine on forest land by 45% from NFI5 to NFI9 (Table 4.8). To a minor extent, peatland drainage had also led to an increase in the mean volume of pine. Between NFI5 and NFI8 the mean volume of spruce increased 17% but remained the same between NFI8 and NFI9. There was a high demand of products made from spruce in 1990s that facilitated the cutting of spruce. The largest relative increase in mean volume by groups of tree species since NFI8 was for the other broadleaved tree species (Tomppo et al. 2001) (Table 4.8). In Figs. 4.20–4.23, the mean volumes of pine, spruce, birch and other broadleaved tree species on combined forest land and poorly productive forest land are presented in map form for both NFI8 and NFI9.

4.7.1.1 Estimates of Mean Volume of Growing Stock by Age Classes and Development Classes

The consequences of intensive forestry can also be seen in the mean volumes by age classes and by tree species (Table 4.9, Table A.12b). Mean volumes are relatively high in young age classes 1–20 and 21–60 years, and had increased particularly in pine dominated forests since NFI8. The mean volumes in age classes 1–60 years had increased significantly in forest dominated by broadleaved tree species. The increase in the mean volume was also high in forests in the 121–160 year age classes. In South Finland, the mean volumes had increased in all age classes compared to NFI8 (Tomppo et al. 2001).

4.7 Growing Stock

Table 4.9 The mean volume by age classes in the 8th and 9th National Forest Inventories

Dominant tree species	Treeless	1–20	21–40	41–60	61–80	81–100	101–120	121–140	141–160	Over 160	Total
	m³/ha										
NFI8											
Pine		11.1	51.0	68.1	95.2	129.5	132.8	119.5	111.2	90.5	74.4
Spruce		20.5	75.7	160.3	188.9	194.9	183.3	150.3	123.5	94.4	142.9
Birch		15.4	65.1	91.5	125.2	127.2	111.8	84.5	79.2	65.2	82.9
Other broadleaved		30.5	98.9	147.0	205.7	170.1	103.0	183.0	123.0	72.0	111.1
Total	2.6	12.9	57.9	93.7	125.3	153.6	149.3	128.7	114.3	92.0	91.8
NFI9											
Pine		17.6	70.8	85.6	99.9	122.0	137.7	133.4	128.1	96.2	85.5
Spruce		19.8	96.7	162.6	204.4	219.4	203.1	178.5	147.6	103.8	150.0
Birch		18.5	78.8	107.5	127.7	138.1	133.4	88.8	37.4	105.1	85.2
Other broadleaved		52.7	128.2	177.4	223.7	206.9	198.0	255.4	–	–	137.5
Total	8.0	18.5	76.2	105.0	130.8	151.3	159.4	145.1	134.4	99.3	100.2

Table 4.10 The mean volume of growing stock by development classes in the 6th and 9th National Forest Inventories on forest land

	Development class									
	1[a]	2	3	4	5	6	7	8[a]	9	All
Inventory	m³/ha									
NFI6	7	11	22	77	140	140	69	–	67	75
NFI9	8	11	24	81	162	174	95	26	81	100

Development classes: *1* temporarily unstocked regeneration stand, *2* young seedling stand, *3* advanced seedling stand, *4* young thinning stand, *5* advanced thinning stand, *6* mature stand, *7* shelter tree stand, *8* seed tree stand in NFI9, *9* low-yielding stand
[a] Class 1 includes also Seed tree stands In NFI6

The mean volume had increased also in all development classes since NFI8. The increase was over 10 m³/ha, e.g., in young thinning and mature stands. The highest mean volume of the development classes was on mature stands, 174 m³/ha (Table A.13a).

Since NFI6 and NFI9, the mean volumes had increased particularly in advanced thinning stands and mature stands (Table 4.10). The proportion of low-yielded stands of forest land decreased from 15% in NFI6 to 11% which partly explains the increase of the growing stock in general (Kuusela 1978) (Table A.13a).

4.7.2 Total Growing Stock Estimates

The growing stock volume on combined forest land and poorly productive forest land was estimated by NFI9 to be 2091 million m³. The sampling error of the growing stock estimate was 11.2 million m³ (0.5%) (Table A.15a). The total growing stock on forest land according to the FAO definition was 2085 million m³ and only 5.3 million m³ on other wooded land (Table A.15d). The area of FAO OWL is small. The volumes of pine and Norway spruce were estimated to be 1,000 and 695 million m³, respectively (Table A.15a). The estimate for pine here includes the volumes of all the other coniferous species except Norway spruce. The total volume of the birch growing stock was 325 million m³. The largest growing stock volumes among the other broadleaved tree species were of European aspen and Grey alder.

The NFI data permitted the calculation of the volume and other characteristics for several land categories. For example, two thirds of the total growing stock was in privately owned forests. Some 75% of the growing stock of spruce was on private land. In North Finland, the proportion of forest area in private ownership was less than in the South (Table A.3): the proportion of the privately owned growing stock was 43% while the largest growing stock was found on state owned land, 47%.

The volume of the growing stock on combined forest and poorly productive forest land on mineral soil was 1,612 million m³, and on peatland soil it was 479 million m³. The growing stock on drained peatlands was 387 million m³ (19% of the total growing stock), of which 51% was pine, 23% spruce and 25% birch (Table A.15b).

The volume of growing stock on forest land and poorly productive forest land available for wood production was 1,963 million m³ (Table A.15c). In North Finland,

4.7 Growing Stock   133

Table 4.11 The volume of growing stock on combined forest land and poorly productive forest land in the 1st to 3rd and 5th and 9th National Forest Inventories

	Field measurement	million m³				
Inventory	Years	Pine	Spruce	Birch	Other broadleaved	Total
(a) Whole country						
NFI1	1921–1924	777	481	290	40	1,588
NFI2	1936–1938	707	502	295	56	1,560
NFI2[a]	1936–1938	624	441	257	48	1,370
NFI3[b]	1951–1953	672	549	282	35	1,538
NFI5[b]	1964–1970	655	555	244	37	1,491
NFI6	1971–1976	686	568	224	42	1,520
NFI7	1977–1984	745	613	249	53	1,660
NFI8	1986–1994	865	691	277	58	1,890
NFI9	1996–2003	1,000	695	325	72	2,091
(b) South Finland in the 8th and 9th National Forest Inventories						
NFI8	1986–1992	509	561	179	47	1,296
NFI9	1996–2000	583	558	206	60	1,408
(c) North Finland in the 8th and 9th National Forest Inventories						
NFI8	1992–1994	356	129	98	11	594
NFI9	2001–2003	416	137	119	12	684

[a] Post-war area
[b] Converted to correspond the volume equations employed since NFI6, yielding 3% increase

where the area of protected land was large, 16% of the total growing stock was not in wood production (see Sect. 4.2).

4.7.2.1 Changes in Growing Stock Estimates Compared to Earlier NFIs

The development of the volume of growing stock is presented for the whole country from NFI1 (1921–1924) to NFI9 and for South and North Finland since NFI8 in Table 4.11. The geographical area of Finland was larger during NFI1 and NFI2 as they included land loss after the World War II. However, the NFI2 estimates are also presented for the post-war geographical area of Finland in Fig. 4.24 and Table 4.11. It should also be noted that different tree volume equations were used in NFI1 and NFI2, for that reason estimates are not fully comparable to other NFIs (Ilvessalo 1942). The volume of the growing stock had increased from 1,538 million m³ in NFI3 to 2,091 million m³ in NFI9, that is, by 36% in a little less than 50 years. The reasons are explained in Sect. 4.7.1. At the same time (1952–2000), the total drain, fellings plus natural losses, were 2,695 million m³, i.e., 1.8 times the growing stock in NFI3. Due to the large area and intentional forest regeneration operations, particularly in North Finland, the volume decreased slightly in 1960s before it began to increase (Table 4.11, Fig. 4.24). The increase in the volume of growing stock has continued since NFI8 when the growing stock was 1,890 million m³. The increase from NFI8 to NFI9 was 9% and 15% in South and North Finland, respectively.

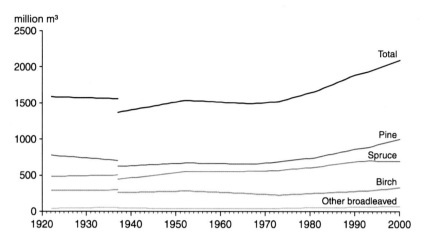

Fig. 4.24 Volume of growing stock on combined forest and poorly productive forest land in the 1st to 9th National Forest Inventories (1921–2003). Note post-war change in geographical area of Finland

Table 4.12 The total growing stock on forest land by diameter classes

Diameter at breast height (cm)					
Inventory	0–9[a] million m³	10–19	20–29	30–	Total
(a) whole country in the 6th to 9th National Forest Inventories					
NFI6	143	480	587	270	1,480
NFI7	182	523	608	310	1,624
NFI8	195	612	639	396	1,843
NFI9	214	718	676	429	2,037
Inventory	million m³				
(b) South Finland in the 8th and 9th National Forest Inventories					
NFI8	119	389	473	304	1,286
NFI9	124	442	493	338	1,397
Inventory	million m³				
(c) North Finland in the 8th and 9th National Forest Inventories					
NFI8	76	223	166	92	557
NFI9	90	275	183	91	640

[a]In NFI6 3–9 cm

4.7.2.2 Changes in Growing Stock Estimates by Diameter Classes

In Table 4.12a and Fig. 4.25, the volumes on forest land by diameter classes for NFI6–NFI9 are presented by four diameter classes of 10 cm intervals with $d_{1.3}$ rounded to the closest integer (Kuusela 1978; Kuusela and Salminen 1991; Tomppo et al. 2001) (Table A.19). The volumes are given by diameter classes separately for

4.7 Growing Stock 135

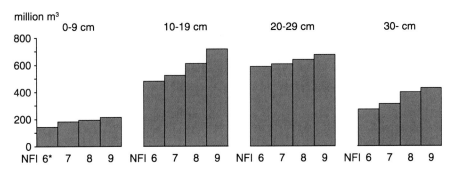

Fig. 4.25 Total growing stock of forest land in diameter classes in the 6th to 9th National Forest Inventories (1971–2003). *) The first diameter class 3–9 cm in the 6th National Forest Inventory

South and North Finland for NFI8 and NFI9 in Tables 4.12b and 4.12c. Since the 1970s (NFI6), the growing stock on forest land had increased continuously in all four diameter classes. The largest relative increase had occurred in the diameter class over 30 cm. The highest absolute increase was in the 10–19 cm diameter class. This was caused by the large area of young forests following the regeneration of mature forests. Since NFI8, the growing stock of all four coarse diameter classes had increased throughout the whole country. The largest increase in growing stock since NFI8 in whole country as well as in South and North Finland was in diameter class 10–19 cm (Table 4.12).

4.7.3 Volume Estimates of Saw-Timber

The production of timber for saw milling has traditionally been an important goal of Finnish silviculture. The estimation of the volume of saw-timber of the sample trees was discussed in Sect. 2.16. Little less than 1/3 of the total growing stock on combined forest land and poorly productive forest land was sawtimber (627 million m^3). The proportions of saw-timber of the total growing stock were 36% and 18%, in South and North Finland, respectively (Table A.16a). The large area of young forests explains the low proportion of saw-timber in North Finland. The proportion of saw-timber in the growing stock was greatest, 33% in privately owned forests. The quality and size requirements of saw-timber had changed between NFIs, also since NFI8. The quality requirements of the timber assortments are derived from the definitions given by the forest industries. The bucking of the sample trees to timber assortments was changed, starting in North Finland in NFI8. Since then, taper curve models have been used and measured lengths of the quality classes. For these reasons, the volumes of saw-timber between the different NFIs are not fully comparable.

On forest land, there were on average 87 saw-timber sized stems (see Sect. 2.16) per hectare and the total number of saw-timber sized stems was 1,762 million (Table A.20a). On forest land available for wood production, the total number of saw-timber sized stems was 1,647 million. The average number of saw-timber sized stems was 118/ha in South and 49/ha in North Finland. The highest number of saw-timber sized stems/ha was in the diameter class 20–24 cm while the highest mean volume of saw-timber sized stems was in the diameter class 25–29 cm (Table A.20a).

4.8 Volume Increment

4.8.1 Increment Estimates

The average annual stem volume increment was 3.8 m^3/ha/year (over bark) in the whole country on forest land and poorly productive forest land (Table A.21a). The volume increment decreases from south to north with the diminishing effective temperature sum (Fig. 1.1), being highest somewhat inland from the southern coastline (Fig. 4.26). In NFI9, the highest averages for forestry centres in the south were between 6 and 7 m^3/ha/year, while in northern Lapland the average increment was 1 m^3/ha/year. The differences in the average annual increment and temperature conditions between forestry centres are illustrated in Fig. 4.27.

The estimated total annual volume increment was 87 million m^3, of which the forests on mineral soils and peatlands accounted for 75% and 25%, respectively (Table A.21b). The proportions of tree species of the total increment varied between mineral soils and peatland soils and between South and North Finland. The proportion of spruce was highest on mineral soils in South Finland and birch species accounted for a large share of the peatland increment. In North Finland, the dominance of pine was clear both on mineral soils and peatlands.

4.8.2 Uncertainties in Increment Estimates and Comparisons with Estimates from Earlier Inventories

The definitions, measurements and estimation methods of volume increment have changed during the decades of forest inventory history in Finland. Climatic variations affecting tree growth have also had an effect on the growth estimates. The uncertainties in comparisons between inventory results are now briefly described before the time series of growth estimates are presented.

4.8.2.1 Sampling Error

The standard error estimate for the total annual volume increment was derived using only increment measurements from sample trees. The estimated relative standard error was 0.7%, which gives a standard error of 0.6 million m^3 for the estimated

4.8 Volume Increment

Fig. 4.26 The average annual increment of the volume of growing stock (m³/ha/year) on combined forest land and poorly productive forest land in the 9th National Forest Inventory

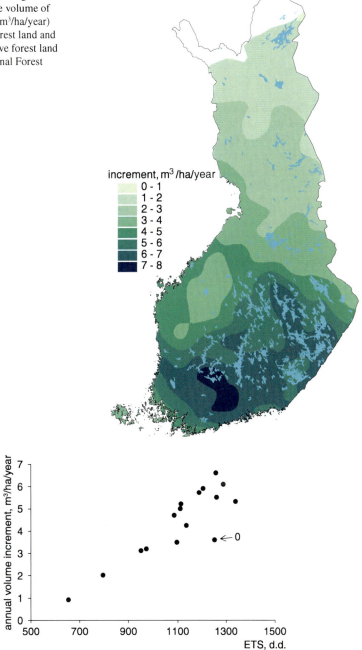

Fig. 4.27 The average annual volume increment (m³/ha/year) in forestry centres as a function of the effective temperature sum (d.d). The average effective temperature sum for the 30-year period 1951–1980 was estimated for each plot (Ojansuu and Henttonen 1983) and the average values for forestry centres were derived from the plot data. '0' indicates the county of Åland

total annual volume increment (86.8 million m³). The samples for increment estimation are independent in different inventories. The sampling error when comparing differences of estimates should thus be below 2 million m³ at a high statistical probability.

4.8.2.2 Changes in Definitions, Measurements, and Estimation Methods

As regards comparisons with earlier inventories, the uncertainty is larger than indicated only by the sampling error. Comparisons with earlier inventories are complicated by:

1. Differences in the target population (e.g. definitions of land classes changed in the 1960s (NFI5), minimum $d_{1.3}$ of measured trees was 2.5 cm before NFI7)
2. Differences in methods of measuring increment cores and height increments
3. Differences in methods of estimating volume increments for sample trees with diameter and height increment measurements (Kujala's (1980) method since NFI6)
4. The increment of drain has been included since the 1960s in NFI5
5. Differences in bark inclusion in increment estimates. The estimates have included the increment of bark since NFI5 and earlier results have been corrected applying average bark percentages
6. Differences in the length of the increment measuring period, which was 10 years in NFI1 and NFI2, and 5 years since NFI3

The target population and estimation of volume increments of sample trees were equal in NFI7–NFI9. Changes have, however, been made in selecting sample trees for increment measurements, boring increment cores, and measuring annual ring widths. In NFI5–NFI7, two increment cores were taken from each sample tree. Since 1990 in NFI8, only one increment core has been taken. In NFI9, changes were made in the methods of measuring increment cores and tree heights. In NFI5–NFI8, all increment cores were measured with a microscope. Since 1999 in NFI9, about 50% of cores have been scanned and ring widths measured using the WinDENDRO™ Software (Regent Instruments Inc., Quebec, Canada). Comparisons between the methods of measuring annual rings were made before introducing the scanning method and no statistically significant differences in ring widths were found (Tuomainen 2000). Changes in height increment caused by new equipment for measuring tree heights were corrected using models (Sect. 2.22).

As regards changes in definitions and increment measurement, the results from NFI7, NFI8, and NFI9 are comparable. In addition, the comparability of NFI6 is fairly good with the exception of trees with diameter below 2.5 cm, which may have some importance in comparing, e.g., increment estimates for birch. Only rough comparisons with the results from e.g. NFI3 can be made.

4.8.2.3 Annual Variation in Increment

Comparison of increment estimates from forest inventories carried out in different years is also complicated by annual growth variation, caused mainly by climatic factors. Other environmental factors causing annual variation in tree-ring widths are, e.g. heavy seed crops and damages caused by insects. This annual variation is usually assessed on the basis of growth indices. The indices express the annual growth levels of trees after standardization, which means removing the effects of most important time-dependent factors affecting growth (e.g. tree age and stand density). For practical reasons, the growth indices are usually estimated for tree ring widths at 1.3 m height. The annual variation in the ring widths at 1.3 m is also reasonably well correlated with the annual variation in the volume growth. Comparisons of inventory growth periods using growth indices have been made in the Finnish national forest inventories since the second inventory in the 1930s (Ilvessalo 1942).

For NFI9, the estimates of growth indices were derived by employing mixed linear models (Henttonen 1990, 2000). The annual growth levels of Scots pine on mineral soils are presented in Fig. 4.28 as an example of annual growth variation. The values in Fig. 4.28 are scaled so that the average of the period 1965–2007 equals 100. For the areas of the forestry centres, the maximum annual deviations from the average level are 20–30%. The use of 5-year periods in the increment estimation reduces the fluctuation in increment estimates. Proceeding from region to region also has a similar effect. However, even for the large areas and time periods of one inventory fieldwork, the differences from the average level may exceed 10%. Table 4.13 shows estimates of the average growth levels in NFI6–NFI9. The estimates in Table 4.13 were derived as averages over South and North Finland without taking into account local differences within regions and their interaction with the progress of the inventory fieldwork.

The diameter increment level in NFI9 was below the average level in South Finland and close to the average level in North Finland. Figure 4.28 also illustrates

Table 4.13 The estimated diameter increment levels in the 6th to 9th National Forest Inventories. The deviations from the average level of the period 1965–2007 in percentage units

Region	Tree species	Diameter increment level in NFI			
		NFI6	NFI7	NFI8	NFI9
South Finland	Scots pine	−7	+13	+2	−8
South Finland	Norway spruce	−7	+6	+5	−3
South Finland	Birches	−10	+15	+11	−10
NorthFinland	Scots pine	−3	+3	−1	+4
North Finland	Norway spruce	−5	−1	−12	0
North Finland	Birches	+8	+3	−5	−1

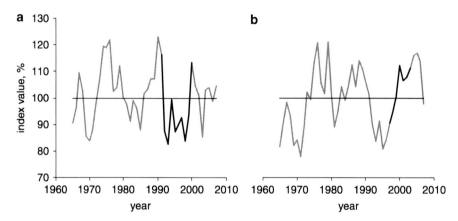

Fig. 4.28 The annual ring indices for Scots pine on mineral soils in South Finland (**a**) and North Finland (**b**). Dark lines indicate the growth measurement years of the 9th National Forest Inventory

the differences between southern and northern Finland. In the south, the inventory period covers a long low-growth period, whereas in the north, the inventory fieldwork started after the turning point towards years of higher growth. Such cyclic variations with long high-growth and slow-growth periods are typical of Scots pine. Compared with both NFI7 and NFI8, the growth level in NFI9 was low in South Finland. In North Finland, differences between inventories are smaller and there are differences between tree species.

4.8.3 Forest Balance

The forest balance enables the compatibility of the volume, drain and increment estimates to be checked. An estimate of the final volume of growing stock is derived by using a volume estimate from an earlier inventory, volume increment estimates and drain statistics. The components of total drain are roundwood removals, logging residues, and natural drain. The drain statistics are compiled by the Finnish Forest Research Institute (www.metla.fi/metinfo/tilasto). This estimate of final volume can then be compared with the volume estimate based on an inventory sample. The forest balance between NFI8 and NFI9 was calculated as (Kuusela and Salminen 1991):

$$initial\ volume + volume\ increment - drain = calculated\ final\ volume$$

where
initial volume = NFI8 estimate of growing stock volume, million m^3,

4.8 Volume Increment

$$\text{volume increment} = \sum_{t_8=1986}^{1994} \sum_{m_8=5}^{11} \sum_{t_9=1996}^{2003} \sum_{m_9=5}^{11} p(t_8, m_8, t_9, m_9) \left(a_{m_8} IV_{t_8} + b_{m_9} IV_{t_9} + \sum_{i=t_8+1}^{t_9-1} IV_i \right)$$

$$\text{drain} = \sum_{t_8=1986}^{1994} \sum_{t_9=1996}^{2003} q(t_8, t_9) \left(0.5 D_{t_8} + 0.5 D_{t_9} + \sum_{i=t_8+1}^{t_9-1} D_i \right),$$

t_8, t_9 are the measurement years in NFI8 and NFI9, respectively,

m_8, m_9 are the measurement months in NFI8 and NFI9, respectively,

$p(t_8, m_8, t_9, m_9)$ is the proportion of the area measured in year t_8 and month m_8 in NFI8 and in year t_9 and month m_9 in NFI9,

$q(t_8, t_9)$ is the proportion of the area measured in year t_8 in NFI8 and in year t_9 in NFI9,

IV_i is the annual volume increment estimate in year i, million m³.

$$IV_i = \begin{cases} \text{NFI9 estimate, if } i > t_9 - 5 \text{ or } i = t_9 - 5 \text{ and } m_9 < 8 \\ \text{linear interpolation between NFI8 and NFI9 estimates, otherwise} \end{cases}$$

a_{m_8} is the proportion of the annual volume increment that accumulated in the measurement year of NFI8 after the measurements made in the month m_8,

$$a_{m_8} = \begin{cases} 0.8, \text{ if } m_8 < 7 \\ 0.3, \text{ if } m_8 < 7 \\ 0, \text{ if } m_8 < 7, \end{cases}$$

b_{m_9} is the proportion of the annual volume increment that had accumulated in the measurement year of NFI9 before the measurements made in the month m_9,

$$b_{m_9} = \begin{cases} 0.2, \text{ if } m_9 < 7 \\ 0.7, \text{ if } m_9 < 7 \\ 1, \text{ if } m_9 < 7 \end{cases}$$

D_i is the annual drain in year i from forest statistics, million m³.

The growth increment of trees accumulates mainly in June and July, which is also a period of active field measurements, which usually start at the end of May. The dates of the field measurements in NFI8 and NFI9 were checked within each forestry centre and the proportion of the increment of NFI8 and NFI9 measurement years (t_8 and t_9) belonging to the period of the balance calculation was included applying monthly coefficients a and b, which were derived from the results of Henttonen et al. (2009).

The length of the balance calculation period varied by regions from 5–6 years in the northern parts of South Finland to 11–12 years in southernmost Finland. Regional differences in the time gaps between inventories make it necessary to calculate the balances separately for each forestry centre. This was also recommended by Kuusela and Salminen (1991). When the forest balance is derived for a long period of time, the number of years with interpolated volume increments increases. The uncertainty caused by annual variations in tree growth increases simultaneously, since the interpolated increments are assumed to be, on average, at the same level of growth variation as the inventory increments. Therefore, the calculations were carried out also with interpolated increments adjusted with growth indices.

The balance for the period between NFI8 and NFI9 is presented in Table 4.14. The difference between the calculated volume of the final growing stock and the volume estimated in NFI9 was 2.4% in South Finland and 1.8% in North Finland, when interpolated increments were adjusted with indices. The difference is higher than in the forest balance calculations of Kuusela and Salminen (1991). In South Finland, the difference between the calculated and estimated volume is high (34 million m^3) even if the sampling error in volume estimates (9 million m^3 in NFI9) is taken into account. The difference between the calculated and estimated volume was overall at the same level in all regions, including the forestry centres with short, 5–6 years, balance calculation period. The results of the forest balance calculations may indicate that the increments are overestimates or drain statistics are biased downwards. The development of sample-based estimation methods for the annual drain is one of the future tasks in the development of national forest inventory in Finland.

Table 4.14 also shows that between NFI8 and NFI9 the ratio of drain to increment has been higher in the south (0.75) than in the north (0.54). The overall ratio was 0.69. According to the results of Kuusela and Salminen (1991), the ratio of drain to increment was 0.71 for the period between NFI6 (1971–1976) and NFI7 (1977–1984).

Example 4.1. An example of the forest balance calculation for a forest centre

The field measurements of NFI8 were made in August 1992 and June 1993. The estimated growing stock volume and annual volume increment were 120.1 million m^3 (= *initial volume*) and 4.4 million m^3/year, respectively.

The field measurements of NFI9 were made in September 2000 and July 2001. The estimated growing stock volume and annual volume increment were 141.9 million m^3 and 6.0 million m^3/year, respectively.

The annual volume increment (IV_i) and drain (D_i) estimates were:

Year $i, i = t_8,...,t_9$	1992	1993	1994	1995	1996	1997	1998	1999	2000	2001
IV_i (million m^3)		5.0	5.2	5.4	6.0	6.0	6.0	6.0	6.0	6.0
D_i (million m^3)	2.9	2.8	3.1	2.8	2.8	3.0	3.3	3.4	3.4	3.5

The volume increments for the period 1993–1995 were interpolated between the NFI8 and NFI9 estimates. In the interpolation, the inventory estimates were set at the years 1990 and 1998, which were the mid-years of the 5-year increment estimation periods in NFI8 and NFI9, respectively. For the year 2001, which is outside the increment estimation period of NFI9, the increment was assumed to be the same as the annual average estimated in NFI9.

4.8 Volume Increment

Table 4.14 The forest balance between the 8th and 9th National Forest Inventories

Region	NFI8 field measurement years	NFI9 field measurement years	NFI8 volume, million m³	Volume increment between NFI8 and NFI9 (million m³)	Drain between NFI8 and NFI9, (million m³)	NFI8 volume + increment - drain (million m³) = A	NFI9 volume (million m³) = B	Difference A-B, million m³ (%)
A. No adjustments for the annual variation in tree growth								
South Finland	1986–1992	1996–2000	1,296	578	432	1,442	1,408	34 (2.4%)
North Finland	1992–1994	2001–2003	594	234	121	707	684	23 (3.3%)
B. Adjusted with growth indices								
South Finland	1986–1992	1996–2000	1,296	578	432	1,442	1,408	34 (2.4%)
North Finland	1992–1994	2001–2003	594	223	121	696	684	12 (1.7%)

The proportions of different periods between the NFI8 and NFI9 field measurements were:

The increment estimates for the periods between NFI8 and NFI9 were:

t_8	m_8	t_9	m_9	$p(t_8, m_8, t_9, m_9)$	$q(t_8, t_9)$
1992	8	2000	9	0.2	0.2
1993	6	2001	7	0.8	0.8

$0+1 \times 6.0 + 5.0 + 5.2 + 5.4 + 4 \times 6.0 = 45.6$ million m³ (NFI8 Aug 1992, NFI9 Sep 2000)

$0.8 \times 5.0 + 0.7 \times 6.0 + 5.2 + 5.4 + 5 \times 6.0 = 48.8$ million m³ (NFI8 Jun 1993, NFI9 Jul 2001)

For the measurements made in Aug 1992 and Sep 2000 in NFI8 and NFI9, respectively, the balance calculation period includes only complete growing seasons ($a_{m8}=0$, $b_{m9}=1$). As regards the period Jun 1993–Jul 2001, the growing season had already started in June and a part of growth had accumulated before the measurements ($a_{m8} = 0.8$). Similarly, in July 2001 only a part of annual growth belonged to the balance calculation period ($b_{m9} = 0.7$). The estimated *volume increment* between NFI8 and NFI9 is a weighted sum of the estimates for the two periods: $0.2 \times 45.6 + 0.8 \times 48.8 = 48.2$ million m³.

The drain estimates for the periods between NFI8 and NFI9 were:

$0.5 \times 2.8 + 2.8 + 3.1 + 2.8 + 2.8 + 3.0 + 3.3 + 3.4 + 0.5 \times 3.4 = 24.3$ million m³ (1992–2000)

$0.5 \times 2.8 + 3.1 + 2.8 + 2.8 + 3.0 + 3.3 + 3.4 + 3.4 + 0.5 \times 3.5 = 25.0$ million m³ (1993–2001)

The estimated *drain* between NFI8 and NFI9 is a weighted sum of the estimates for the two periods: $0.2 \times 24.3 + 0.8 \times 25.0 = 24.9$ million m³.

Finally, the *calculated final volume* for the time of NFI9 is thus 143.4 million m³ (= 120.1 + 48.2 − 24.9). The growing stock volume estimate in NFI9 was 141.9 million m³.

4.8.4 Changes in Annual Volume Increment Estimates since the 1950s

Volume increment estimates from NFI3-NFI9 are presented in Fig. 4.29 by tree species. Figure 4.30 shows the total annual volume increment estimates together with the total annual drain. As mentioned in Sect. 4.8.2, the changes in measurements and volume increment estimation allow only rough comparisons with inventories before NFI6 and NFI9. Figure 4.29 nevertheless indicates that the volume increment of pine began to increase already in the 1960s, and that the increase has continued between NFI8 and NFI9 in spite of a period of low tree level growth in South Finland (Table 4.15).

4.8 Volume Increment

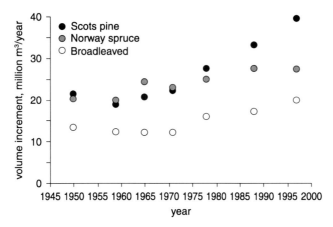

Fig. 4.29 The annual total volume increment estimates from the 3rd to the 9th National Forest Inventory by tree species

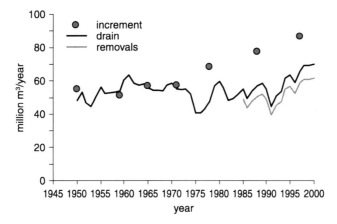

Fig. 4.30 The annual total volume increment estimates from the 3rd to the 9th National Forest Inventories, total drain in 1950–1999, and roundwood removals in 1985–1999

For Norway spruce, the increase in volume growth is not as steady as for Scots pine. The volume increment estimates of Norway spruce were slightly lower in NFI9 than in NFI8, which was probably caused by the low-growth years at the end of 1990s in South Finland, which is the main region for spruce growth.

The comparison of increment and drain series in Fig. 4.30 shows that a rapid increase in total volume increment began in the 1970s as the volume of growing stock started to increase and annual drain remained low compared to the increment.

On mineral soils, the increase in the increment is mainly the result of the changes in the density and structure of the forests (increases in the average increment). The estimated area of mineral soils was only 200,000 ha (1%) larger in NFI9 than in the end of the 1960s (NFI5). This increase results mainly from the re-classification of

Table 4.15 The estimates of the annual volume increment of growing stock in the 8th and 9th National Forest Inventories

Region	Inventory	Field measurement years	Scots pine	Norway spruce	Birches	Other broadleaved species	Total
			million m³/year				
SouthFinland	NFI8	1986–1992	21.4	24.1	9.3	3.6	58.5
North Finland	NFI8	1992–1994	11.6	3.3	3.8	0.5	19.2
Whole country	NFI8	1986–1994	33.1	27.5	13.1	4.1	77.7
South Finland	NFI9	1996–2000	23.3	22.9	10.2	3.9	60.3
North Finland	NFI9	2001–2003	16.2	4.4	5.3	0.6	26.6
Whole country	NFI9	1996–2003	39.6	27.3	15.5	4.5	86.8

drained peatlands into mineral soils (Sect. 4.4.1). The land-use changes between forestry, agriculture and other land-use classes have slightly decreased the forest area on mineral soils (Statistics Finland 2010). On the peatlands, drainage has also increased the forest area and improved the productivity of sites.

4.8.4.1 Changes in Annual Volume Increment by Age Classes on Mineral Soils

Changes in silvicultural practices have lead to higher growing stock (Sect. 4.7) and these have apparently been decisive factors for the increase in the average increment on mineral soils. In Fig. 4.31, the NFI7 and NFI9 estimates of average increment (m³/ha/year) and growing stock volume (m³/ha) on mineral soils in South Finland are presented by age class. Increments in this figure include only living trees, because the present estimation method of the increment of drain does not allow allocation of the drain increment to plots or strata deviating from forestry centres. In NFI9, the increment of drain was 4.4 million m³ (5% of the total volume increment). Growing stock volumes and volume increments were higher in all age classes in NFI9 compared with NFI7 (years 1977–1982 in South Finland). The rapid early development of young planted and seeded stands can also be seen, especially in the 21–40 year age class, which also covers large areas according to NFI9 (Tables A.12 and A.23).

4.8.4.2 Changes in Annual Volume Increment Estimates on Peatland Soil Forests Since the 1950s and the Effect of Peatland Forestry on the Increment

When assessing the changes of the volume increment of forests on peatland soils between inventories, it is necessary to take into account that the area of peatland varies between the inventories to some extent due to the differences in assessments, in addition to the changes in the definitions and methods (Sect. 4.8.2). For instance,

4.8 Volume Increment

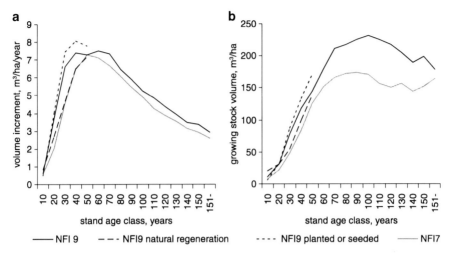

Fig. 4.31 The estimates of the average annual volume increment (excluding the increment of drain) (**a**) and growing stock volume (**b**) by age classes in South Finland in the 9th National Forest Inventory (1996–2000) and the 7th National Forest Inventory (1977–1982)

the area of peatland was 150,000 ha higher in NFI9 than in NFI8. A likely reason is that site assessment, particularly distinguishing of mineral soils and peatland soil, was more thorough in NFI9 than in NFI8. Another reason for the area changes between inventories is the moisture condition that affect the ground vegetation composition and judgment of the site between mineral soils and peatland soil. A further aspect that should be considered when comparing results to inventories earlier than NFI8, is the fact that the area of peatland soils has decreased due to drainage operations. Peatland soils with a thin peat layer tend to loose the peat layer and the ground vegetation may also change to resemble that of mineral soils, wherefore the site should be classified into mineral soils. The change between NFI3 and NFI9 was assessed to be 510,000 ha (Sect. 4.4.1).

The question at issue is, what is the share of the increase of the volume increment between NFI3 and NFI9, on one hand, on forests which were classified as mineral soil forests in NFI3, and on the other hand, on forests which were classified as peatland forests in NFI3. In addition to the current site classifications, the conversions between mineral soils and peatland soils have to be taken into account. The conversion from mineral soils to peatland soils can be assumed negligible. For the area converted from peatland to mineral soils, the estimate of 510,000 ha can be used (Sect. 4.4.1). It can be assumed that this converted areas were drained mineral soils in NFI9. The estimated average annual volume increment on drained mineral soils was 4.7 m^3/ha/year in NFI9. It is thus estimated that the volume increment of the forests that were peatland in NFI3 but were classified as mineral soils in NFI9 was 2.40 million m^3/year. The total increase in annual volume increment of the peatland forests between NFI3 and NFI9 would therefore be 14.1 million m^3 (23.6–9.5, Table 4.16) and for mineral soil forests 17.5 million m^3. However, the changes in the

Table 4.16 The annual volume increment on mineral soils and peatland soils (million m³) on productive forest land and forest land of poor growth in the 3rd National Forest Inventory (1951–1953) and on forest land and poorly productive forest land in the 8th (1986–1994) and in the 9th National Forest Inventories (1996–2004)

Inventory	Field measurement years	Mineral soils	Peatland	Peatland 1951–1953	Total
NFI3	1951–1953	45.7	9.5	9.5	55.2
NFI6	1971–1976	47.2	10.2	–	57.4
NFI7	1977–1984	53.5	14.9	–	68.4
NFI8	1986–1994	60.3	17.4	20.1	77.7
NFI9	1996–2003	65.6	21.2	23.6	86.8
Change (%)	NFI3–NFI9	43.5	123.2	148.4	57.2

In the column "Peatland 1951–1953", the increments have been assessed for all those sites which were peatland in the 3rd National Forest Inventory but were either peatland or drained mineral soils in the 8th and 9th National Forest Inventory respectively

definitions between NFI3 and NFI9 should be noted. One important issue when assessing the volume increment changes on peatland forests is to recall that the trees with $d_{1.3}$ smaller than 2.5 cm have not been measured in NFI3. The number and volume increment of those trees are relatively high on some peatland sites. The increment of the drain is also missing in the NFI3 estimates. It is also possible that cuttings on peatland forests have been at a lower level than in mineral soil forests, which partly explain the higher increase in the volume increment (Paavilainen and Tiihonen 1988).

It was assessed in the national forestry programs before the intensive peatland drainage period that the drainage area could be 6.5–7.5 million ha. Heikurainen (1961) estimated that the increase of the annual volume increment of the peatland forests with a drainage area of 6.5 million ha would range from 12 to 14 million m³/year. Heikurainen (1982) further increased the assessment slightly, ending up to the total increment of peatland forests up to 18–23 million m³/year. This assessment seems to be relatively accurate although the drained area is lower than assumed in the calculations.

4.9 Protected Areas

Nature conservation in Finland has been realized through diverse types of protected areas. Threats to vulnerable ecosystems and scenic values have been the main aspects of nature conservation. National parks, strict nature reserves, mire conservation areas, protected herb-rich forest and old-growth forests have legal status together with associated regulations. Most of the conservation areas were established on state-owned land but the state also acquired land for protection. Nowadays, protected areas are established increasingly on private land. Habitat and species conservation, as well the maintenance and enhancement of biodiversity, has also extended nature protection to managed forests. The protected areas discussed in this section cover

4.9 Protected Areas

Table 4.17 The total area, protected area and the proportion of protected area by principal site classes and site fertility classes on mineral soils on forest land, poorly productive forest land and unproductive land. The site fertility classes refer to Table 2.8

Mineral soils Principal site class	Site fertility	Total area 1,000 ha	Protected area	Protected of the total area %
	1	386	24	6.2
	2	2,281	82	3.6
	3	7,603	611	8.0
	4	4,271	381	8.9
	5	482	75	15.6
	6	8	2	25.9
	7	689	180	26.2
	8	1,383	1,098	79.4
	Total	17,101	2,452	14.3
Spruce mires		2,370	192	8.1
Pine mires		5,069	405	8.0
Open bogs and fens		1,623	575	35.4

all lands reserved for nature conservation on forest land, poorly productive forest land and unproductive land.

According to NFI9, there were 3.6 million ha of protected areas in Finland which was 12% of the total land area. Most protected areas were quite small and sparsely located in South Finland but in North Finland they created large areal units. Because much land in the north is state-owned, 88% of all protected areas were in North Finland, and 91% were state-owned (Table A.3).

The aim in nature conservation is to preserve representative examples of the all nature habitat types in Finland. One way to take stock of the conservation situation is to compare how much of the area of different habitat types is protected. NFI classifies mineral soils into site fertility classes that represent also a coarse classification of habitats. Because the majority of the large national parks and wilderness reserves are in the north, the infertile and barren sites were well represented of which almost 80% were protected (Table 4.17). Of the more fertile sites, less than 10% were conserved. Of the open bogs and fens, 35% were located in the conservation areas whereas of the peatlands with tree cover, about 8% were protected.

From the perspective of the growing stock volume, the implications of forest protection for wood production are not significant. Only 7% of Finland's total growing stock volume 2,091 million m^3 was in protected forest land and poorly productive forest land: 16% in the north and 3% in the south.

Results concerning protected areas of forest land, protected and non-protected together, are presented by age classes at 20 years intervals, and all forest older than 160 years in all in Tables A.12a–c. Using the same classification for protected forest land, over half of the area in the north was in the over 160 years class. To clarify how old the oldest forests were, forest stands were re-classified into 40-year age classes with the uppermost class placed at over 320 years. As Fig. 4.32 shows, in North

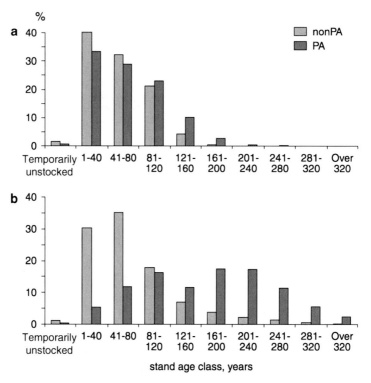

Fig. 4.32 The age class distribution of protected forest land (PA) and forest land outside protected areas (nonPA) in South Finland (**a**) and North Finland (**b**)

Finland the old forests were more likely to be protected areas than the younger forests. Almost 40% of the protected forests were over 200 years old. In South Finland, the age class structure of protected forest land was quite similar to normal forests (Fig. 4.32).

4.10 Forest Damage

In the NFI field measurements, forest damage was assessed from stands belonging to forest land according to the national definition. Information about forest damages was recorded using four variables: symptom description, age of the damage, causal agent, and degree of the damage. The defoliation of coniferous sample trees was also assessed.

The degree of damage categorized the severity of a disturbance using four classes. In the assessment of the degree, the main criteria were the impact of the damage on the growth, mortality and timber quality of the growing stock, and on the productivity of the forest stand. Occasional symptoms of damages, such as few fallen and

4.10 Forest Damage

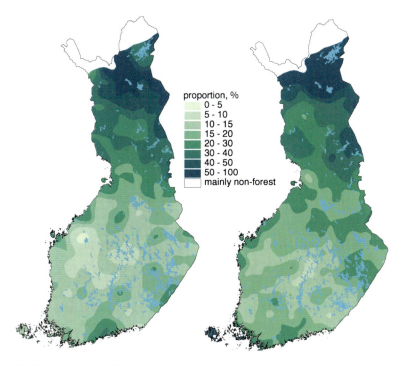

Fig. 4.33 The proportion of forest land where significant (moderate, severe or complete) damages where registered in the 8th and 9th National Forest Inventories

broken trees, are often apparent, so it is often difficult to determine whether or not the damage has any impact on the vitality or quality of the growing stock, and thus, whether it should be described or not. Because the NFI field data are widely used in research, information on forest damages, even slight symptoms of recognizable damaging agents, can be interesting, for example, when studying the relation of specific damaging agents to other forest variables, and in research concerning their temporal pattern (Nevalainen 1999, 2002; Nevalainen and Yli-Kojola 2000). Consequently, slight damages that had no direct affect on the quality of a stand were also described. The symptoms and the causal agent of the most significant damage were described where there were symptoms of several disturbances. However, the joint effect of all damage on the stand was taken into account when defining the degree of the damage.

In the whole country, forest damage was observed on 47% of the forest land area (Table A.34). Most of the damage was slight (47% of the area affected by damage). Moderate or severe damage comprised 41% and 11% of the damage area respectively (Table A.34, Fig. 4.33). Complete damage, requiring the regeneration of a stand, occurred on 72,800 ha (1% of the affected area). Compared to NFI8 (1986–1994) the area of forest damage had increased. In NFI8, significant damage was present on 22% of the forest land (Yli-Kojola and Nevalainen 2006). Damage was significant if the degree of damage was moderate or severe, or if the stand was totally destroyed, and consequently, lowered the silvicultural quality of the stand. In NFI9, damages affecting the stand quality occurred on 25% of the forest land area.

Damage was more frequent and it was often most sever in North Finland. Significant damage was found on 34% of the forest land area in North Finland, and on 17% in South Finland (Fig. 4.33). It should be noticed that damage was assessed in both wood production and protected forests (Sect. 4.2). Forest management recommendations stress the prevention of forest damage, for example, when tending seedling stands and during regular thinnings to maintain the vitality of the growing stock (Uotila and Kankaanhuhta 1999). Furthermore, when harvesting, trees with a broken top or stem deformation are removed, and signs of damages cleaned in this way. In protected forest, where neither cuttings nor silvicultural measures are carried out, more symptoms can be observed and damages occur more frequently. The area of protected forests is greater in North than in South Finland (Sect. 4.2) which explains the higher amount of forest damages there.

The causal agent responsible for the damage was identified on 58% of the damaged areas in the whole country, and the reason for the observed damage remained unknown for the remaining area (Table A.34). Uncertainty about the causal agent was more common in case of slight damages, and in North Finland also for more severe damages. The large amount of unknown causes of damage (50%) in North Finland is partly explained by the symptom description 'multiple symptoms'. In practice, such stands were over-aged and gradually degenerating, and there were several damage symptoms and causal agents present. In the case of 'multiple symptoms', causal agents were not recorded, and in the results (Table A.34) the area of these multiple damages is classed as an unknown agent. Thus, in North Finland, the area of unknown agents is actually somewhat lower, since multiple symptoms were observed on about 4% of the damage area.

Abiotic damages that decreased the quality of a stand occurred on over 1 million ha (5.4% of the forest land area). The extent of abiotic damages was at the same level as in NFI8 (5%) (Yli-Kojola and Nevalainen 2006). For NFI9, categories of abiotic causal agents were specified to include, in addition to wind and snow, separate classes for frost, fire and soil factors. After snow damages, soil factors, e.g. drought, flood, soil frost and deficiency or imbalance of nutrients, were the second most frequent abiotic causal agents. Most often such damage was moderate (Table A.34). Forest fires are quite rare in Finland due to efficient fire prevention measures. The climatic conditions and mosaic of peatlands also slows the spread of fire. Burnt-over areas amount to 530 ha/year on average during the period of NFI9 (Finnish Statistical Yearbook of Forestry 2006). Damage caused by forest fires was found on 30,900 ha, occurring mainly in North Finland.

During NFI9, storms with a wind force of over 20 m/s were unusually frequent, and some devastating local tornadoes occurred (Finnish Meteorological Institute 2010). In November 2001, two storms caused windfalls on 1.2 million ha of forest land in southern Finland (Ihalainen and Ahola 2003). The NFI9 sample plots were measured in that region during the period 1996–1999, and to estimate the extent of the damaged area and loss of timber volume, a special damage inventory of windfall

4.10 Forest Damage

trees was carried out by remeasuring the permanent sample plots shortly after the storms in winter 2001 (Ihalainen and Ahola 2003). Wind throw trees must be harvested in order to avoid secondary damages e.g. infections of insects and fungi, and once the fallen and broken trees cleared away, there are no observable damages. Thus, neither all wind damage nor other damage could be seen in the results of NFI9 if damaged trees had already been removed. Wind damage occurred on 1% of the forest land area (236,800 ha) in the whole country (Table A.34), whereas on 2% in NFI8 (Yli-Kojola and Nevalainen 2006).

The most common symptoms of damage were stem deformations that cover all kind of stem curves and boughs caused by, for example, earlier top damages (Table A.35). In North Finland, this is related to heavy snowfall: the crown snow load breaks tree tops and branches especially at high altitudes. Consequently, the largest group of identified causal agents in North Finland was abiotic, present on 12% of the forest land area, snow being the most frequent agent of all. Conversely, in South Finland, a distinct reason for stem deformation and other top damage was not found as several damaging agents cause similar defects. The most common group of identified causal agents in South Finland was fungi, for some fungal disease was present on 10% of forest land area. However, the most frequent single agent was moose (*Alces alces* L.). Moose browse on young tree shoots and twigs, preferable broadleaved trees but also pine, especially in winter time. Symptoms of earlier moose browsing remain as stem deformation even when the risk of browsing is over. During NFI9 (in 1996–2003) the density of the moose population was relatively high (about 3 moose per 1,000 ha) compared, for example, to the period of NFI8 (Finnish Statistical Yearbook of Forestry 2006). This can be seen as increased areas of moose browsing damages. In NFI8, moose damages were on 2%, and in NFI9 on 3% of the forest land area (Table A.34; Yli-Kojola and Nevalainen 2006).

In Scots pine (*Pinus sylvestris* L.) dominated stands, slight damages were more frequent than severer damages, while in Norway spruce (*Picea abies* (L.) Karst.) and broadleaved dominated stands, more significant damages were often present. One explanation could be that the typical damage in Norway spruce stands, for example spring frost in young stands and wind throw in mature stands, is usually drastic once it occurs. In the case of root rot (*Heterobasidion annosum*), the symptoms of damage are observed only when the disease is already well advanced.

According to the defoliation assessment of conifers, needle loss was more common in the case of Norway spruce than Scots pine (Table A.36). In the whole country, 17% of assessed Norway spruces, and 8% of Scots pines suffered slight needle loss, defoliation being over 25%. Defoliation increased by the tree age, and was more severe in North Finland, which is consistent with the results of previous inventories (Lindgren et al. 1998; Yli-Kojola and Nevalainen 2006). The assessment of defoliation was first included in NFI8 (1986–1994), as it produces information on forest vitality and there was initially considerable concern about forest decline. Since then, the causes for temporal and spatial variation of defoliation have been widely studied, for example within the ICP Forests Programme (International Cooperative Programme for Monitoring and Assessment of Air Pollution Effects on

Forests). It has been stated that the die-back of trees, expressed as defoliation, is a complex phenomenon that reflects the fluctuating effects of abiotic and biotic factors acting simultaneously or consecutively. Therefore, any trend in defoliation is difficult to detect (Nevalainen and Yli-Kojola 2000).

4.11 Silvicultural Quality of Forests

4.11.1 Silvicultural Quality

Silvicultural quality of a forest stand was assessed from the wood production point of view, and the evaluation was based on the principles of good management practices (see Sect. 2.13). For example, the presence of standing dead wood, old trees or some uncommercial tree species decreased the stand quality although they are important and intended features for forest biodiversity. Furthermore, as protected forests are set-aside from timber production, the results of stand quality are presented only for forest land available for wood production.

Although silvicultural quality is an assessed characteristic and the results vulnerable to subjective interpretations, the areas in question, particularly the areas of the low-yielding stands, provide important information for forest policy and when allocating public subsidiaries to forestry. Even a special law has been passed to reduce the area of low-yielding stands (Laki Lapin 1982). To determine the silvicultural status of forests for both forest policy and planning the operations at regional level is the main reason to assess the silvicultural quality in NFIs.

The silvicultural quality of the forests was considered to be good in 34% of forests in the country as a whole (Fig. 4.34, Table A.22). In general, the silvicultural quality

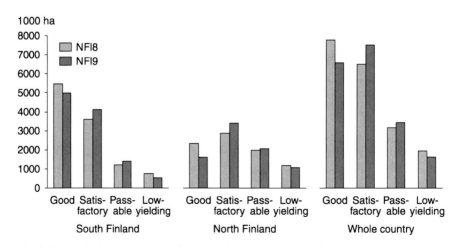

Fig. 4.34 The silvicultural quality of stands on forest land available for wood production in South and North Finland in the 8th and 9th National Forest Inventories

4.11 Silvicultural Quality of Forests

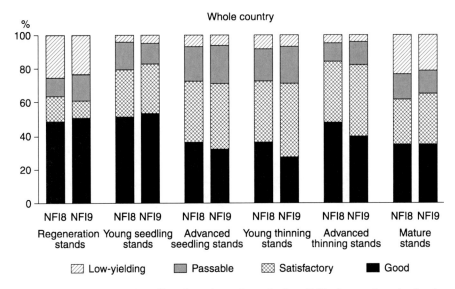

Fig. 4.35 The silvicultural quality of stands on forest land available for wood production by development classes in the 8th and 9th National Forest Inventories

was better in South Finland than in North Finland. The area of low-yielding forests was 1.6 million ha (8%) of the forests. Compared to NFI8, the area of forests with good or satisfactory silvicultural quality (73%) was nearly same but the area of low-yielding forests had decreased by 330,000 ha (Tomppo et al. 2001).

The most common reason for the decrease in the stand quality was uneven structure, i.e. an unfavorable spatial distribution of the growing stock. This typically caused only a slight decrease in the silvicultural quality because the amount of growing stock (basal area or number of seedlings) could be sufficient but the poor spatial distribution (clusters and openings) or unwanted variation of tree size and age reduced the yield. The reasons for low-yielding quality were somewhat different in South and North Finland. In South Finland, the most common reason was species composition (24%), which means that unfavorable tree species decrease both yield, and technical quality and value of the growing stock. In North Finland, the most common reason was the over-aged growing stock (22%). However, compared to the results of NFI8, the area of low-yielding stands caused by both these reasons had decreased and, for example, the area of low-yielding stands caused by poor technical quality and forest damage had increased (Tomppo et al. 2001).

In regeneration forests (i.e. temporarily unstocked, seed tree and shelterwood stands), the silvicultural quality was good in 50%, and low-yielding in 23% of the areas (Fig. 4.35, Table A.22). The total area of low-yielding regeneration stands was 130,000 ha. In these stands, the time since clear cutting was over 4 years and the establishment of a new stock had not been carried out, or in the case of natural regeneration, it could be seen that regeneration was not going to succeed in the appropriate time. The proportion of low-yielding regeneration areas was higher in

North Finland than in the South. Especially natural regeneration seemed to fail more often in North Finland. Compared to NFI8, the area of low quality regeneration stands (i.e. passable or low-yielding) had decreased, although their proportion of the total regeneration area had increased (Fig. 4.35).

The quality of young seedling stands was mostly good or satisfactory (84%), and only 5% were low-yielding (Fig. 4.35, Table A.22). The quality of advanced seedling stands was slightly lower: the proportions were 71% and 6%, respectively. Furthermore, the proportion of passable stands was rather high (23%). The passable quality indicates that the number of seedlings was too low and complementary planting was recommended if permitted by the mean height of the growing stock. If regeneration had failed or a complete damage had occurred, a stand was classified as low-yielding and had to be re-established on the basis of good practice. The most common reasons for decreased quality in seedling stands were uneven spatial distribution of seedlings and damage.

In young thinning stands, the silvicultural quality was good or satisfactory in 72%, and in advanced thinning stands in 82% of the area. The proportions of low-yielding stands were low, 7% and 4%, respectively (Fig. 4.35, Table A.22). The reasons for low-yielding class in young thinning stands were mainly species composition, natural under-density and poor technical quality.

The silvicultural quality of mature stands was good or satisfactory in 65%, and low-yielding in 21% of the area (Fig. 4.35, Table A.22). The major reason for low-yielding quality was over-aged growing stock which caused reduced growth, damage and mortality. In this development class, the differences between South and North Finland were the most evident. In North Finland, 42% of the mature forests were assessed as low-yielding, while in South Finland the figure was only 8%. The proportions of quality classes were similar to those in NFI8, but the area of low-yielding mature stands had decreased by 220,000 ha (Tomppo et al. 2001). However, this was not only because of the improvement in stand quality but also due to the decrease of the total area of mature forests available for wood production. For example, a new nature conservation programme of old-growth forests was approved at the beginning of the 1990s.

4.11.2 Methods and Success of Regeneration

In addition to establishment type of a stand, i.e. natural seed generated, natural sprout generated, planted or seeded, the success of the regeneration was assessed in the field. For the description of the growing stock in seedling stands, the number of seedlings capable for further development and the number of all seedlings were counted on circular sample plots located 20 m apart around the centre of the NFI plot. The radius of the sample plots was 2.3 m. The seedlings capable for further development included seedlings that were vital, of suitable species and were of a suitable size and position in the stand, i.e. the seedlings that were left after tending or thinning the seedling stand.

The area of naturally established stands in the whole country was 7.2 million ha, and the area of artificially regenerated stands (i.e. planted or seeded) 4.6 million ha, which is 39% of the total area of the seedling and young thinning forests (Table A.23). The proportion of artificially regenerated stands had increased 4%-units since NFI8 (Tomppo et al. 2001). The area of failed artificial regeneration was 310,000 ha, or 7% of the total area of the artificially regenerated stands. The regeneration was considered to have failed if the number of seedlings planted or seeded was less than the lower limit (trees/ha) in the passable quality class. The silvicultural quality was better in artificially regenerated than in naturally regenerated stands. Of the artificially regenerated stands, 38% were good and 23% passable or low-yielding, whereas in naturally regenerated stands, proportions were 24% and 37%, respectively.

In the young coniferous seedling stands, about 16% of the area contained too few seedlings capable for further development (Table A.24). In these stands, the density was less than 1,500 seedlings/ha, which is close to the limit for recommending complementary planting. From the advanced seedling stands dominated by pine, 31% of the area were fairly sparse (the density was under 1,500/ha). The figure was 23% for stands dominated by spruce. In the young seedling stands dominated by broadleaved trees, 18% of the area were too sparse (density under 1,000 seedlings/ha), and for the figure was 25% for the advanced seedling stands. On the basis of the figures in Table A.24, the success of regeneration between coniferous and broadleaved trees cannot be compared because there are coniferous stands with failed artificial regeneration among the sparse broadleaved seedling stands, for example, failed stands planted with pine or spruce that have regenerated naturally with birch.

4.12 Management Activities

4.12.1 Accomplished and Proposed Cuttings

The area estimates concerning the accomplished cuttings, silvicultural measures and drainage operations are based on NFI9. The Forest Statistics unit of the Finnish Forest Research Institute collects similar annual information (Finnish Statistical Yearbook of Forestry 2005). This information is based on reports on forestry operations by the forestry centres, forest industry companies and private forest owners. In NFI9, the proposed measures were based on the silvicultural state of each individual forest stand and on the current forest management practice recommendations (Metsätalouden kehittämiskeskus Tapio 1994). These assessments did not take into account the sustainability of wood production. This should be remembered, particularly in the case of regeneration cuttings. The long-term sustainability of forestry is taken into account when making the long-term cutting scenarios (Siitonen et al. 1996; Nuutinen et al. 2005; MELA 2010). The results of the proposed measures in NFI9 are presented only for forest land available for wood production, even though in the field they were assessed on all sample plot stands regardless of the protection status of the forest. As with the silvicultural quality of the stands, the accomplished, and particularly the proposed cutting and silvicultural measures are important information for forest policy, in preparing forestry programs and in allocating public subsidiaries to forestry.

During the 10-year period prior to NFI9, different cutting operations (including tending of seedling stands) had been carried out on 6.2 million ha, that is on 31% of the forest land area (Table A.25). The area of stands on which the tending of seedlings had been accomplished was 1.5 million ha, which was clearly less (by 560,000 ha) than in NFI8 (Tomppo et al. 2001). The area of regeneration cuttings achieved during the previous 10-year period was 1.5 million ha which corresponded to regeneration cuttings on 0.75% of the forest land annually, on average. Compared to NFI8, the area of cuttings for artificial regeneration had remained as before but the area of cuttings for natural regeneration had increased by 120,000 ha (Tomppo et al. 2001). Furthermore, the area of thinnings, especially of first thinnings, had clearly increased (Tomppo et al. 2001).

The area of forests where no cuttings were observed, or the time since the most resent cutting was more than 30 years, was 28% of the forest land area in the whole country (Table A.27). The proportion of un-cut areas was higher in North Finland (42%) than in South Finland (16%). The reason for this is the greater amount of protected forests in North Finland. A total of 4.5 million ha (24%) of forest land available for wood production in the whole country had not been cut during the past 30 years.

On the poorly productive forest land, cuttings were accomplished on 2.2% of the area during the previous 10-year period before the inventory (Table A.27). No cuttings were observed, or the time since the latest cutting was over 30 years, on 86% of the poorly productive forest land, and on 81% of the poorly productive forest land available for wood production.

Proposals for cuttings were made for the 10-year period following the inventory. In the field, a thinning was usually proposed only if the basal area of the growing stock before cutting at the proposed time was at least 6 m^2/ha higher than the thinning limit of the thinning regimes (Metsätalouden kehittämiskeskus Tapio 1994). For mature stands, regeneration cutting could be proposed when the minimum age or diameter for regeneration was reached (Metsätalouden kehittämiskeskus Tapio 1994). The timing for the proposed cutting was recorded as delayed if the silvicultural quality of the stand had decreased in such a way that the yield had reduced because of the delay of the cutting.

For the 10-year period following the inventory, cuttings were proposed for 60% of the forest area available for wood production (Table A.26). From this area of 11.5 million ha, 21% (2.4 million ha) were proposals for the tending of seedling stands. The area of proposed first thinning (3.1 million ha) was 2.7 times, and the area of regeneration cuttings (3.3 million ha) 2.1 times higher than the area of accomplished cuttings (Fig. 4.36). The proportion of natural regeneration from the proposed regeneration cuttings and from the accomplished cuttings were both around 30%. However, the area of proposed cuttings, for example regeneration cuttings of 326,000 ha/a, is only the sum of the silvicultural recommendations not a management goal for the 10-year period following the inventory. According to the

4.12 Management Activities

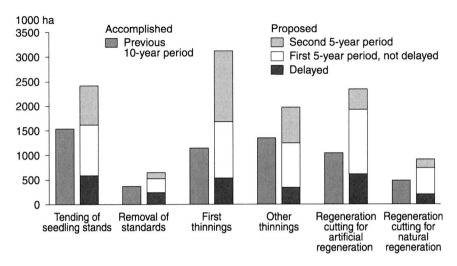

Fig. 4.36 The area of accomplished and proposed cutting operations by cutting type and timing on forest land available for wood production

analysis of forest production and utilization possibilities based on the NFI9 data, the area of regeneration cuttings for the subsequent 10-year period was 215,000 ha/a in the cutting scenario of maximum sustainable removal (Nuutinen et al. 2005). Compared to NFI8, the area of proposed thinnings had increased, and the area of regeneration cuttings had decreased, especially with respect to natural regeneration (Tomppo et al. 2001).

The timing of the proposed cutting was assessed to be delayed on 22% of the proposals (2.6 million ha) (Table A.26). First and other thinning were assessed to be less frequently delayed than other cuttings (Fig. 4.36). The area of stands where the proposed cutting was already delayed had increased since NFI8 (Tomppo et al. 2001). The delayed tending of seedlings had increased by 150,000 ha, delayed thinnings by 270,000 ha, and delayed regeneration cuttings by 430,000 ha (Table A.26, see also Tables A.13a and b; Tomppo et al. 2001).

4.12.2 Accomplished and Proposed Silvicultural Measures

The silvicultural measures considered were planting or seeding, complementary planting or seeding, and pruning (Table A.28). During the 10-year period prior to the inventory, planting or seeding was accomplished on 1.1 million ha forest land. This was higher than the area of accomplished cuttings for artificial regeneration in the whole country, and especially in North Finland (Table A.25). One reason for this was the planting or seeding on areas where, for example, natural regeneration had been unsuccessful. The area of accomplished complementary planting was 33,000 ha, and the area of pruning 160,000 ha. However, it is likely that some complementary planting accomplished over 5 years ago could not be detected in the field.

The proposed silvicultural measures for the subsequent 10-year period included planting or seeding, complementary planting, weeding of a seedling stand, and clearing of a regeneration area (Table A.29). Complementary planting was proposed only if the number of seedlings was below the recommended amount for a satisfactory seedling stand. Clearing was proposed only for natural regeneration areas where existing vegetation was hindering the regeneration, or for low-yielding stands where no commercial timber was obtainable. The area of proposed planting or seeding was 500,000 ha (Table A.29). About half of the area concerned open regeneration forests, and the rest low-yielding seedling stands or stands where natural regeneration had failed. The area of proposed complementary planting was 56,000 ha.

The area of accomplished soil preparations during the previous 10-year period was 1.3 million ha (Table A.30). The area had increased compared to NFI8 as had the area of regeneration cuttings (Tomppo et al. 2001). In addition, the area was higher than the planted or seeded area (1.1 million ha) because soil preparation measures had been carried out also on natural regeneration areas. The proportions of light soil preparation methods (harrowing and scarification) and mounding had increased, and the proportion of ploughing decreased. For example, during the 5-year period prior to the inventory, the area of ploughing was 5%, whereas during the 11–30 year period before the inventory it had been 35%. In general, heavier soil preparation methods had been employed in the North because of thicker humus layers found there.

The area proposed for soil preparation at the current stage of the forest stand was 500,000 ha (Table A.31), which was 150,000 ha more than in NFI8 (Tomppo et al. 2001). One reason for the increased area could be that during the 10-year period prior to the inventory the area of regeneration cuttings was 235,000 ha higher than the area of accomplished soil preparation measures. In South Finland, the proportion of light soil preparation methods was 84%, and the proportion of mounding 16%, which corresponds to the proportions of accomplished methods. In North Finland, the proportions of proposed heavy preparation methods, ploughing (15%) and mounding (24%), were higher than in South Finland, and furthermore, slightly higher than their proportions of accomplished methods in preceding 10-year period.

4.12.3 Drainage Operations

The forest drainage and other operations related to soil water balance were recorded on forest land, and on peatlands also on poor productive forest land and unproductive land. The observation period for accomplished drainage operations was 30 years prior to the inventory. During the 10-year period prior to the inventory, forest drainage operations were carried out on 1.0 million ha (Table A.32). Of these operations, 75% was on peatlands and the rest on wet mineral soils. From the area of

4.12 Management Activities

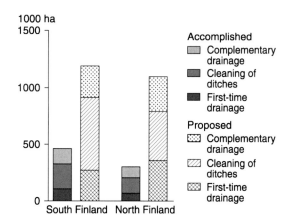

Fig. 4.37 Area of the accomplished drainage during the 10 year period prior to the inventory and proposed drainage on peat land in South and North Finland

9.1 million ha of peatland on forestry land in Finland, 4.9 million ha has been drained (see Sect. 4.4). The major part of first-time drainage on peatlands was done between 1950s and 1980s, consequently, accomplished operations in NFI9 were mainly ditch cleaning and complementary draining (Fig. 4.37). The area of first-time drainage was 170,000 ha which was more than 50% less than in NFI8 (Tomppo et al. 2001). However, the total area of draining operations on peatlands had increased because of extensive ditch cleaning and complementary draining in South Finland.

Drainage operations on wet mineral soils were carried out on 240,000 ha, of which almost 70% was first-time drainage (Table A.32). In South Finland, the area had increased and had more than doubled compared to the area in NFI8 (Tomppo et al. 2001). In North Finland, the area of drainage operations on wet mineral soils had slightly decreased, but the proportion of ditch cleaning and complementary draining had increased.

The area drained for other than for forestry purposes was 19,000 ha during the previous 10-year period (Table A.32). These drainage operations included, for example, road ditches and agricultural ditches, and they were considered only if they had an effect on the growth of the growing stock, or when the drainage system included the whole stand. NFI9 found only a small area of peatlands on which mire restoration had been carried out (1,400 ha). The blocking of ditches to restore drained mires to their natural state only began at the beginning of 1990s (Kaakinen and Salminen 2006).

The area of proposed drainage describes the future need of draining operations from the silvicultural point of view and the suitability of the site for timber production (Table A.33). The suitability was defined according to the peatland type, effective temperature sum and quality of the existing growing stock (age, density, technical quality and revival capability). The area of peatland suitable for first-time forest drainage on forest and poorly productive forest land available for wood production was 630,000 ha (Fig. 4.37, Table A.33). If this area was drained, the remaining area of undrained peatland would be 3.5 million ha, of which 300,000 ha would be forest land. However, first-time drainage on peatlands of natural state are no longer carried out (PEFC Finland 2010).

Ditch cleaning or complementary drainage was proposed for 1.7 million ha (34%) of drained peatlands (Fig. 4.37, Table A.33). The proposed area was about three times higher than the area of accomplished ditch cleaning and complementary drainage during the previous 10-year period (Table A.32). In general, the area of both proposed and accomplished ditch cleaning and complementary drainage had increased since NFI8 (Tomppo et al. 2001).

The area of forest land on wet mineral soils, i.e. mineral soils paludified, was 820,000 ha which of 580,000 ha was undrained. First-time forest drainage was proposed for 310,000 ha, and ditch cleaning or complementary draining for 99,000 ha of wet mineral soils (Table A.33). On drained mineral soils, the need for ditch cleaning or complementary drainage was much lower (7% of the drained area) than on drained peatlands (34%). The total area of proposed drainage operations on wet mineral soils was 71,000 ha lower than in NFI8 (Tomppo et al. 2001).

4.13 Biodiversity Indicators

4.13.1 Biodiversity Measurements in NFI

Field data collected in NFIs have traditionally described some environmental factors that have an influence on biodiversity, sometimes, biodiversity itself. Examples are site fertility, soil type, tree species composition by crown layer and, in some inventories, understory vegetation. For NFI9, some new biodiversity measurements were added. The most important of these were the inventory of forest key habitats, dead wood and keystone tree species, the results of which are presented in this section.

Many of the results presented previously in this chapter also carry information on forest biodiversity. For example, forest site types, forest age, development stages, tree species composition of growing stock and forest damages all depict aspects of forest biodiversity and its extent.

The COST Action E43 framework (COST Action E43 2010), sought to develop methods for harmonising forest resource, forest carbon pool and forest biodiversity estimation. Twenty-seven European countries, the Forest Inventory and Analysis Programme (FIA) of the U.S. Forest Service and Scion from New Zealand participated in the Action, in addition to other institutions and organisations that closely followed the work. Measures of the amount and quality of dead wood were chosen to be one of the seven key variable groups that describe forest biodiversity in different vegetation zones and are possible to be measured in large area forest inventories. The other variables or variable groups were forest type, forest structure (species composition, horizontal and vertical structure), forest age, composition of ground vegetation, naturalness of the habitat, and regeneration (Tomppo et al. 2010).

4.13.2 Key Habitats

Key habitats – here forest habitats with a high biodiversity value, mainly as defined in the Finnish Forest Act (1996) – were classified into 33 classes in NFI9. Observations were made on the naturalness of the key habitat, and its ecological value as well as on forestry operations performed on the site (see Sect. 2.15). All observations on key habitats were recorded no matter what their state of naturalness. However, the state of naturalness and the ecological value determined whether an individual habitat was a Forest Act habitat (i.e. "a habitat of particular significance" defined in the Finnish legislation), and therefore protected by the Finnish Forest Act, or a key habitat without legislative protection. In this section, areas of the habitat classes are presented by their ecological value. Furthermore, areas of 13 ecological habitat groups derived from the individual habitat classes are considered.

A total of 4.8 million ha (18%) of forest land, poorly productive forest land and unproductive land in the whole country was categorized as key habitat (Table A.41, see Sect. 4.2 for the criteria for the protected areas). Of this area, 1.2 million ha were on land that was protected or planned to be protected. A total of 460,000 ha of the habitats met the Forest Act habitats criteria except possibly the criteria of a small size. Thus, the greater part (4.3 million ha) of the area of the key habitats did not meet the Forest Act criteria. At least a part of this area can be considered to have potential as high biodiversity sites should the management regimes of these sites favour the restoration and maintenance of the habitats. On the other hand, a considerable area of peatland habitats had been drained, and the costs of restoration might be very high in comparison to the expected results. Some of the valuable peatlands were too large to meet the Forest Act criteria.

The area of Forest Act habitats corresponded to 1.7% of the area of forest land, poorly productive forest land and unproductive land. Of the Forest Act habitats, 36% were found inside protected areas (Table A.41). The most abundant Forest Act habitats were open fens and bogs (20% of the area of the Forest Act habitats) open rocks (17%), and mesotrophic hardwood-spruce mires (12%), as well as stone- and boulderfields, ombrotrophic pine bogs and *Sphagnum fuscum* dominated bogs. Of these, only hardwood-spruce mires were of significance to forestry. There were some differences between South and North Finland in the relative abundance of the different Forest Act habitats. For example, hardwood-spruce mires were the most abundant habitats in South, while open fens and bogs and ombrotrophic pine bogs were the most abundant in the North. Note that the relative sampling errors of the area estimates of individual habitat classes were rather large where the area estimate was small.

Rather than of comparing single habitat classes, it might be more informative to look at habitat groups. Key habitats can be grouped into 13 distinctive ecological habitat groups that differ in many ways regarding e.g. moisture and trophic conditions, habitat size and characteristics of growing stock. The habitat groups were:

1. Springs (and their surroundings of) (includes key habitat classes 1,2, see Table 2.23)
2. Brooks (and their surroundings of) (3)

3. Small ponds (and their surroundings of) (4, 5)
4. Eutrophic spruce mires (7, A)
5. Eutrophic fens (8, 9, D)
6. Herb-rich forest patches (G, H, J, K, L, M)
7. Islets of mineral soil forest surrounded by un-drained peatlands (P)
8. Gorges and ravines (and their surroundings of) (R, S)
9. Precipices (and their surroundings of) (T)
10. Sand fields (Y)
11. Rocks and stonefields (U, W, X)
12. Oligo- and ombrotrophic mires with only sparse tree stands (I, B, V, C, E)
13. Alluvial meadows (F)
14. Others (6, N, Z).

The most abundant Forest Act habitat groups in both South and North Finland were oligo- and ombrotrophic mires (37% of the area of the Forest Act habitats in South, 28% in North Finland), the second most were rocks and stonefields in South (35%) and eutrophic spruce mires in North Finland (23%) (Table 4.18). Valuable key habitats not meeting the Forest Act criteria were also dominated by oligo- and ombrotrophic mires both in South and especially in North Finland. In South Finland, most of the not valuable key habitats were herb rich forests and eutrophic spruce mires, whereas in North Finland oligo- and ombrotrophic mires were most abundant also in this value class.

About one half (48%) of the area of all key habitats was at a natural state (Table A.42). The habitats where a large proportion of the areas was at a natural state were mostly those that were not suitable for wood production, e.g., eutrophic and mesotrophic fens (83% of the area at the natural state), fen or bog surrounding a small pond, open fens and bogs (78%) as well as stonefields and boulder fields (80% of each were at a natural state). On the other hand, a high proportion of the areas of the habitats suitable for wood production were often strongly altered. Examples were eutrophic paludified hardwood-spruce forest, mesotrophic hardwood-spruce mire, of which respectively 56% were strongly altered, and of the different herb-rich forests, 40–50% were strongly altered.

Forestry operations had not been carried out during the 30 years preceding the inventory in about two thirds of the key habitat areas (3.2 million ha) (Table A.43). The greater part of the area of the untreated habitats were habitat types with low wood production values. On a little less than one third of the area of the habitats (1.4 million ha), the habitat was not taken into account when carrying out forestry measures. These were typically habitats suitable for wood production, for example different herb-rich forests.

Key habitats had been taken into account in forestry measures on 9% of the key habitat area in South and on 2% in North Finland; whereas measures for enhancing the value of the habitat had been carried out on 0.3% of key habitats in the South and 0.0% in the North. However, no operations had been carried out in 33% of the key habitats or in their surrounding stands for at least 30 years in the South and 78% in the North. The remaining 58% of key habitats in South and 21% in North Finland had not been considered in forestry measures.

4.13 Biodiversity Indicators

Table 4.18 Areas of key habitat groups (1,000 ha) by habitat value in South and North Finland in the 9th National Forest Inventory

Group	Not valuable	Valuable	Forest Act	Total
South Finland				
Springs (surr)	1.2	1	0.6	2.9
Brooks (surr)	5.5	4.6	3.5	13.6
Ponds (surr)	3.8	6	6.4	16.3
Eutrophic spruce mires	211	29.4	10.3	250.7
Eutrophic fens	2.7	2	3.2	7.8
Herb-rich forests	208.9	57	19.2	285.1
Islets of mineral soil	0.2	0.3	0.6	1.1
Gorges and ravines (surr)	0	0	0	0.1
Precipices (surr)	4.8	6.2	4.2	15.2
Sand fields	0	0	0.1	0.2
Rocks and stonefields	34.8	58.3	76.4	169.4
Oligo- and ombrotrophic mires	101.5	149.7	80.8	332.1
Alluvial meadows	3.3	11.3	10.9	25.5
Others	2.3	6.5	1.9	10.8
Total	580	332.3	218.2	1,130.6
North Finland				
Springs (surr)	1	2.2	2	5.2
Brooks (surr)	2.4	9	15.9	27.3
Ponds (surr)	2.3	20.1	20.1	42.5
Eutrophic spruce mires	120.9	90.3	54.2	265.4
Eutrophic fens	43	153.6	8.3	204.8
Herb-rich forests	15.2	15.4	11.8	42.4
Islets of mineral soil	5.1	7.6	7.4	20.1
Gorges and ravines (surr)	0	0.3	1	1.3
Precipices (surr)	0	0.1	1.6	1.8
Sand fields	0.6	0	2.1	2.7
Rocks and stonefields	14.4	103.6	33.1	151.1
Oligo- and ombrotrophic mires	572.2	1,756.1	65.9	2,394.2
Alluvial meadows	4.3	14	5.1	23.4
Others	108.8	332.5	10.6	451.9
Total	890	2,504.9	239.2	3,634.1

4.13.3 Dead Wood

4.13.3.1 The Volume and Quality of Dead Wood

Dead wood, especially coarse dead wood, has a considerable and widely recognised role in forest ecosystem functions (e.g. Esseen et al. 1997; Jonsson et al. 2005; Siitonen 2001). Coarse dead wood not only hosts a wide range of species but also serves as a carbon reserve and is one of the five carbon pools in greenhouse gas reporting (Penman et al. 2003). Information on the amount and quality of dead wood is required for management and monitoring purposes at different levels.

The mean volume of dead wood (diameter ≥10 cm) on forest and poorly productive forest land was 2.8 m³/ha in South and 8.3 m³/ha in North Finland (Table A.37, Table 4.19). The highest mean volumes were found in North East Finland and the lowest in West Finland (Ostrobothnia) (Fig. 4.38). Overall, a gradient from the South West to North East could be partly explained by forest management history. Standing dead trees made up a little less than one third of the volume of dead wood in South Finland (the mean volume was 0.8 m³/ha). The proportion of standing dead wood in North Finland was over 20% (the mean volume was 1.8 m³/ha). Factors affecting the differences include the intensity of forest management and the age structure and tree species composition of the forests, as well as the slower decaying rate of dead wood in the North.

The mean volume of dead wood on forest land only was, in South Finland, similar to that on the combined forest land and poorly productive forest land due to the fact that the share of poorly productive forest land is small in the South (Table 4.19, Table A.1). The mean volumes on forest land were higher in North Finland, particularly on mineral soils. One reason for the difference was the higher mean volumes of the growing stock on forest land than on poorly productive forest land.

The mean volumes of dead wood were higher on mineral soils than on peatlands, also due to the lower volumes of growing stock on peatlands and the fact that some drained peatlands have not hosted a previous tree generation (Table 4.19). On forest land, the differences between mineral and peatland soils were higher in the North than in the South and actually quite small in South Finland, but on poorly productive forest land, the differences were clear both in the South and North.

There were some differences in the tree species composition of dead wood between South and North Finland. The most abundant species was pine (*Pinus sylvestris*) with a 40% contribution to the total volume in South Finland and 64% contribution in the North (Fig. 4.39, Table A.37). The second most abundant species was spruce (*Picea abies*) (27% in the South and 16% in the North) and the third was birch (*Betula pendula* and *Betula pubescens*) (14% in the South and 12% in the North). This reflects the species composition of the growing stock: in South Finland 41% of the growing stock volume was pine, 40% spruce and 15% birch, and in North Finland 61% pine, 20% spruce and 17% birch (see Table A.15). Aspen (*Populus tremula*) and other broadleaved species were quite scarce in South Finland and very scarce in the North: this also reflects differences in the volume of the growing stock. Overall, coniferous species accounted for 70% of the total dead wood volume in South Finland and 84% in the North, while broadleaved species accounted for 24% in the South and 13% in the North.

The tree species composition of standing and lying dead wood differ very little; only the broadleaved species other than birch and aspen were more abundant as standing than lying trees. Unidentified tree species were naturally more abundant in the lying tree class. In South Finland, the species left unidentified, or identified only as being broadleaved or coniferous due to advanced stages of decay, contributed to 10% of the total dead wood volume. The corresponding figure for North Finland was 8% (Table A.37, Fig. 4.39.)

4.13 Biodiversity Indicators

Table 4.19 The mean volumes of standing and lying dead wood (m³/ha) on mineral soils and peatlands on forest land and poorly productive forest land

	Standing dead wood			Lying dead wood			Total		
	Mineral soil	Peat land	Total	Mineral soil	Peat land	Total	Mineral soil	Peat land	Total
	m³/ha								
South Finland									
Forest land	0.8	0.7	0.8	2.1	1.7	2.0	3.0	2.4	2.8
Poorly productive forest land	1.0	0.4	0.6	1.7	0.4	0.9	2.7	0.8	1.5
Forest land and poorly productive forest land	0.8	0.7	0.8	2.1	1.5	2.0	3.0	2.2	2.8
North Finland									
Forest land	2.3	1.0	1.9	9.5	2.4	7.6	11.8	3.4	9.5
Poorly productive forest land	1.7	1.4	1.5	3.2	1.2	1.8	5.0	2.6	3.4
Forest land and poorly productive forest land	2.2	1.2	1.8	8.9	1.9	6.4	11.1	3.1	8.3

Fig. 4.38 The average volume (m³/ha) of standing and lying dead wood, as defined in the 9th National Forest Inventory, on combined forest land and poorly productive forest land

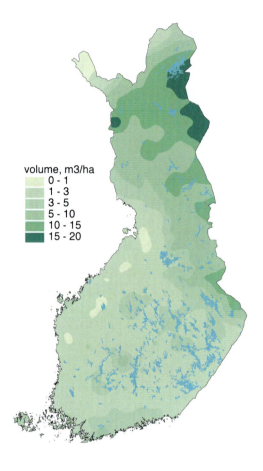

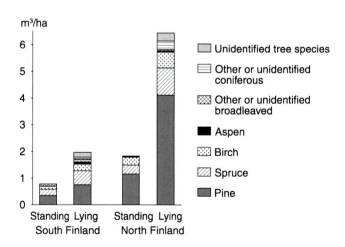

Fig. 4.39 The mean volumes of both standing and lying dead wood by tree species on forest and poorly productive forest land for South and North Finland

4.13 Biodiversity Indicators

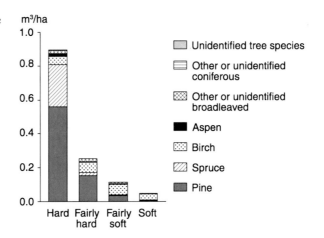

Fig. 4.40 The mean volume of standing dead wood by decay classes and by tree species on forest and poorly productive forest land

Large dead wood is important especially for some insects, polypores and other wood-inhabiting fungal species (Siitonen 2001). The volume of dead wood with a diameter of ≥30 cm, was 0.4 m^3/ha in South Finland and 1.8 m^3/ha in the North (Table A.38). This amounts to 15% of the total dead wood volume in the South and 22% in the North. The greater part of the volume of the large dead trees consisted of pine or spruce. Both in South and North Finland about three quarters of the volume of large dead wood was accounted for by lying dead trees.

The most abundant physical appearance class for standing dead wood was the whole standing dead tree (22% of total volume in South and 18% in North Finland). The second commonest class was the broken standing tree (Table A.39). For lying dead wood, the most abundant physical appearance classes were the uprooted lying tree (24% of total volume in the South and 29% in the North) and the broken lying tree class. The division into physical appearance classes was quite similar in South and in North Finland. Man-made dead wood (stumps or snags, butt ends or bolts and logging waste) covered 18% of the total dead wood volume in the South and 13% in the North.

Most of the dead wood volume was hard (*decay class* 1) in South Finland (35% of volume) but soft (*class* 4) in the North (26% of volume) (Table A.40). In South Finland, the second most abundant class was soft (19%), and in North Finland, hard (21%). The greater part of the volume of standing trees had not decayed, hard material being dominant (Fig. 4.40), but the volume of the lying trees was spread evenly over all decay stages, the soft class being the most abundant class both South and North Finland (Fig. 4.41). Standing dead coniferous trees were mostly hard, whereas standing broadleaved trees showed more advanced decayed. This is self-evident since a more advanced decayed tree trunk is less likely to remain standing than a less decayed one. The distribution of the volume of lying trees into the decay classes was more even especially in South Finland. The volume distribution was skewed towards decay classes 3–5 (soft, rather soft and very soft) with some differences between the tree species. The contribution of the very soft class, the most thoroughly decayed class, was 15% of the total volume (20% of lying tree volume) in South and 18% (24% of lying tree volume) in the North.

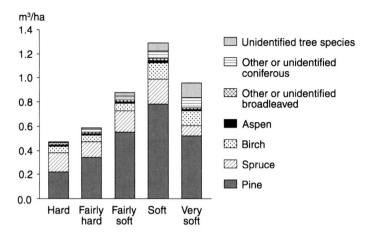

Fig. 4.41 The mean volume of lying dead wood by decay classes and by tree species on forest and poorly productive forest land

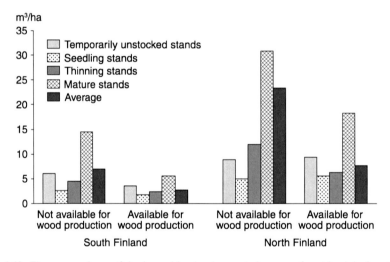

Fig. 4.42 The mean volume of dead wood by development classes on forest land, in forests not available and available for wood production for South and North Finland

4.13.3.2 Dead Wood in Forests Not Available for Wood Production

In NFI9, forestry land was divided into forests that are available for wood production and forests that are not because their use is restricted by some standards (see Sect. 4.2). The volume of dead wood was clearly greater in forests not available for wood production. In South Finland, the mean volume of dead wood on forest land was 2.7 m^3/ha in forests available for wood production and 7.1 m^3/ha in other forests. In North Finland, the difference was even greater, 7.7 m^3/ha in forests available for wood production and 23.3 m^3/ha in other forests (Fig. 4.42). On poorly productive

4.13 Biodiversity Indicators

forest land, the differences were less pronounced both in South and in North Finland due to lower volumes of growing stocks and less intensive management practices on the poorly productive forest land than on forest land.

In mature forests in South Finland (Table 2.17), the mean volumes of dead wood were 5.5 m^3/ha in forests available for wood production and 14.5 m^3/ha in forests not available, the figures for North Finland being 18.2 and 30.9 m^3/ha respectively (Fig. 4.42). The management history of forests outside wood production varies and, for example, some protected areas include forests in early development classes. The difference in the volume of dead wood between forests available and forests not available for wood production was either small or non-existent in development classes other than mature stands.

In forests available for wood production, the volume of standing dead wood varied between stands of different development classes. The peak was in mature stands whereas the lowest point was in seedling stands as the trees from the previous tree generation had probably fallen down. Also the volume of lying dead wood varied, being at its greatest in mature stands and open regeneration areas and lowest in seedling and thinning stands as the former trunks and branches had decayed away.

4.13.3.3 The Volume of Usable Dead Wood

Although all dead wood had not been measured in NFIs before NFI9, the volume of usable dead wood (wood still hard enough to be used, at least as fuel wood) was also measured in many of the earlier inventories. In NFI9, in addition to dead wood measurements on fixed-size circular sample plots, usable dead wood was measured separately as a part of tally tree measurements (see Sect. 2.16). However, it should be noted that usable dead wood was also included in the dead wood measurements and results presented in Sect. 4.13.3. The volume of usable dead wood had been increasing since NFI7 (1977–1984), the first inventory comparable in this sense, especially in South Finland (Fig. 4.43). The average amount of usable dead wood was at its highest in mature stands whereas seedling stands contained the least usable dead wood, both in the South and North (Table 4.20).

4.13.4 Keystone Tree Species

Tree species and individuals that are regarded to be very important for forest biodiversity, referred to as keystone tree species, were inventoried from a fixed radius plot. The tree individuals were selected and measured using species-specific minimum thresholds for breast height diameter (see Sect. 2.18). The estimate of the mean stem number on forest land or poorly productive forest land of keystone tree species was 5.5 stems/ha in the South and 1.5 stems/ha in the North (Table A.44). The most abundant keystone tree species in South Finland was Black alder (*Alnus glutinosa*) (1.9 stems/ha) and Goat willow (*Salix caprea*) in the North (1.2 stems/ha).

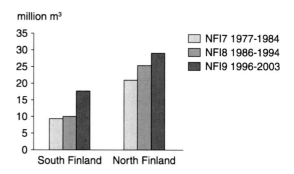

Fig. 4.43 The volume of usable dead wood in South and North Finland in the years 1977–2007

The next most abundant keystone tree species in South were Goat willow (1.6 stems/ha), European mountain ash (*Sorbus aucuparia*) (0.8 stems/ha), European aspen (*Populus tremula*) (0.6 stems/ha) and Grey alder (*Alnus incana*) (0.3 stems/ha), and European aspen in the North (0.2 stems/ha).

However, it is important to bear in mind that the minimum thresholds for breast height diameter varied between species, the largest being 30 cm for European aspen and 20 cm for Grey alder, whereas the lowest was 5 cm for several species. The natural distribution area also varies between tree species; the distribution area for many broadleaved species is limited to the narrow Hemiboreal forest vegetation zone in the south-western corner of the country.

4.13.5 Retention Trees

The description of retention trees was part of the biodiversity monitoring introduced in NFI9. In the 1990s, the renewed forest legislation (Forest Act 1996) and the concomitant recommendations for forest management practices emphasized the maintenance of biodiversity also in managed forests (Metsätalouden kehittämiskeskus Tapio 1994, 2001). For example, in connection with regeneration cuttings, it was recommended that standing dead trees and snags, as well as individual or groups of living trees, are left un-cut in order to produce dead wood, and so provide adequate habitats for flora and fauna. In NFI9, the new variables '*retention tree classes*' (Table 2.14) and '*abundance of retention trees*' sought to collect information on how the new guidelines had been applied in practice. The two variables described what kind of trees were left, their quantity and how they were distributed in a stand. The assessments were on a nominal scale.

In addition to these two new variables, a retention tree storey was recorded as a separate *tree storey* (Table 2.16) in the description of the growing stock. In NFI9, retention tree storeys were recorded on 1.2% of the forest land area (Table A.14). This does not, however, include all stands with retention trees because the retention tree storey was described separately only if the number of retention trees was at least 10–30/ha, whereas in the assessment of '*retention tree classes*' the lowest abundance class was 1–4 trees/ha.

4.13 Biodiversity Indicators

Table 4.20 The mean volume of usable dead wood (m³/ha) by development classes on forest land and on poorly productive forest land

	Forest land								Poorly productive forest land	Forest land and poorly productive forest land
	Temporarily unstocked stands	Young seedling stands	Advanced seedling stands	Young thinning stands	Advanced thinning stands	Mature stands	Shelter or seed tree stands	Total		
	m³/ha									
South Finland	1.0	0.5	0.3	1.1	2.2	3.1	1.4	1.5	1.2	1.5
North Finland	2.8	0.9	0.4	1.1	2.2	8.8	2.1	2.6	2.2	2.5

The '*retention tree classes*' included not only individual, usually large, mature trees but also patches of natural seedlings, usually under-growth of the previous tree generation, left after regeneration cuttings. However, natural seedlings are generally left for regeneration purposes rather than for biodiversity, as they can be used to complement the planting. The "buffer zones of trees" and "whole stands" classes could also have confused the assessment, especially with regard to which tree individuals were interpreted as retention trees in the field. Furthermore, retention trees were found in all development classes, not only on regeneration or seedling stands. This was expected as standing dead trees and snags have commonly been left in all cuttings as nesting trees for birds. Furthermore, seeding trees on a natural regeneration area may have been left, unintentionally or on purpose, for further growth, and former seed trees could therefore also be found as an upper-storey in thinning stands. Thus, in regeneration cuttings retention trees could not be differentiated as being left intentionally and specifically for biodiversity purposes when collecting the field data.

Both new variables were on a nominal scale, while the classifications tried to capture all possible occurrences of retention trees. However, information on the volume of retention trees could not be obtained. Because the interpretation of retention trees was found challenging and the result subjective, the measurements were reconsidered and further developed for NFI10. The retention tree variables were omitted in NFI10, and instead, the retention trees have been measured as part of the tree measurements on the angle count sample plot. The variable crown layer indicates the retention trees.

References

Aaltonen VT (1935) Zur Stratigraphie des Podsolprofils besonders von Standpunkt der Bodenfruchtbarkeit. I. Selostus: Valaisua podsolimaan kerrallisuuteen silmälläpitäen varsinkin maan viljavuutta I. Communicationes Instituti Forestalis Fenniae 20(6):1–150

Aaltonen VT (1939) Zur Stratigraphie des Podzolprofils besonders von Standpunkt der Bodenfruchtbarkeit. II. Communicationes Instituti Forestalis Fenniae 27(4):1–133

Aaltonen VT (1941) Zur Stratigraphie des Podzolprofils besonders von Standpunkt der Bodenfruchtbarkeit. III. Communicationes Instituti Forestalis Fenniae 29(7):1–47

Aaltonen VT (1947) Studien über die Bodenbildung in den Hainwäldern Finnlands mit einige Boebachtungen über ausländische Braunerden. Communicationes Instituti Forestalis Fenniae 35(1):1–92

Aaltonen VT (1951) Soil formation and soil types. In: Suomi, A general handbook on the geography of Finland. Kustannusosakeyhtiö Otava, Helsinki, pp 65–73

COST Action E43 (2010) Harmonisation of national forest inventories in Europe: Techniques for common reporting. http://www.metla.fi/eu/cost/e43/. Accessed 15 Nov 2010

Esseen PA, Ehnström B, Ericson L, Sjöberg K (1997) Boreal forests. Ecol Bull 46:16–47

FAO-UNESCO (1988) Soil map of the world. Revised Legend. World Soil Resources Report 60. Rome, FAO, 119 p

FFCS (2006) Finnish forest certification system. http://www.ffcs-finland.org/. Accessed 14 Dec 2010

Finnish Meteorological Institute (2010) Suomen myrskyt. Ilmatieteen laitos. http://www.fmi.fi/saa/tilastot_21.html. Accessed 20 Nov 2010 (In Finnish)

References

Finnish Statistical Yearbook of Forestry (2005) Finnish Forest Research Institute, pp 103–151. http://www.metla.fi/metinfo/tilasto/julkaisut/vsk/index.html. Accessed 20 December 2010 (In Finnish)

Finnish Statistical Yearbook of Forestry (2006) Finnish Forest Research Institute. http://www.metla.fi/metinfo/tilasto/julkaisut/vsk/index.html. Accessed 20 December 2010 (In Finnish)

Forest Act (1996) Forest Act. 1093/1996. http://www.finlex.fi/en/laki/kaannokset/1996/en19961093. Accessed 1 Mar 2010

Glinka K (1914) Die Typen der Bodenbildung, ihre Klassifikation und geographische Verbreitung. Verlag, Berlin, 365 p

Heikurainen L (1961) Metsäojituksen vaikutuksesta puuston kasvuun ja poistumaan hakkuusuunnitteiden laskemista varten (Summary: the influence of forest drainage on growth and removal in Finland). Acta Forestalia Fennica 71(8):1–71

Heikurainen L (1982) Peatland forestry. In: Laine J (ed) Peatlands and their utilization in Finland. Finnish Peatland Society, Finnish National Committee of the International Peat Society, Helsinki, pp 53–62. ISBN 951-99402-9-4

Henttonen H (1990) Kuusen rinnankorkeusläpimitan kasvun vaihtelu Etelä-Suomessa. Summary: variation in the diameter growth of Norway spruce in southern Finland. University of Helsinki, Department of Forest Mensuration and Management, Research Notes 25, Helsinki, 88 p

Henttonen H (2000) Growth variation. In: Mälkönen E (ed) Forest condition in a changing environment – the Finnish case. Kluwer Academic, Dordrecht, pp 25–32. ISBN 0-7923-6228-4

Henttonen HM, Mäkinen H, Nöjd P (2009) Seasonal dynamics of the radial increment of Scots pine and Norway spruce in the southern and middle boreal zones in Finland. Can J For Res 39:606–618

Hirvelä H, Nuutinen T, Salminen O (1998) Valtakunnan metsien 9. inventointiin perustuvat hakkuumahdollisuusarviot vuosille 1997–2026 Etelä-Pohjanmaan metsäkeskuksen alueella. Julkaisussa: Etelä-Pohjanmaa. Metsävarat 1968–97, hakkuumahdollisuudet 1997–2026. Metsätieteen aikakauskirja – Folia Forestalia 2B:279–291 (In Finnish)

Hökkä H, Kaunisto S, Korhonen KT, Päivänen J, Reinikainen A, Tomppo E (2002) Suomen suometsät 1951–94. Metsätieteen aikakauskirja 2B:201–357 (In Finnish)

Ihalainen A, Ahola A (2003) Pyry-ja Janika-myrskyjen aiheuttamat puuston tuhot. Metsätieteen aikakauskirja 3:385–401 (In Finnish)

Ilvessalo Y (1927) The forests of Suomi Finland. Results of the general survey of the forests of the country carried out during the years 1921–1924. Communicationes ex Instituto Quaestionum Forestalium Finlandie 11, 421 p + Tables 192 p (In Finnish with English summary)

Ilvessalo Y (1942) Suomen metsävarat ja metsien tila: II valtakunnan metsien arviointi (The forest resources and the condition of the forests of Finland: the Second National Forest Survey). Communicationes Instituti Forestalis Fenniae 30:1–446

Ilvessalo Y (1956) Suomen metsät vuosista 1921–24 vuosiin 1951–53. Kolmen valtakunnan metsien inventointiin perustuva tutkimus (The forests of Finland from 1921–24 to 1951–53. A survey based on three national forest inventories). Communicationes Instituti Forestalis Fenniae 47(1):1–227

Ilvessalo Y (1960) The forests of Finland in the light of maps. Communicationes Instituti Forestalis Fenniae 52(2):1–70 (In Finnish with English summary)

IUSS Working Group WRB (2006) World reference base for soil resources 2006. A framework for international classification, correlation and communication, World Soil Resources Reports 103. FAO, Rome. 128 p. ISBN 92-5-105511-4. http://www.fao.org/ag/Agl/agll/wrb/doc/wrb2006final.pdf. Accessed 20 December 2010

Jonsson BG, Kruys N, Ranius T (2005) Ecology of species living on dead wood – lessons for dead wood management. Silva Fennica 39:289–309

Kaakinen E, Salminen P (2006) Mire conservation and its short history in Finland. In: Lindholm T, Heikkilä R (eds.) Finland – land of mires, vol 23, The Finnish Environment. The Finnish Environment Institute, Helsinki, pp 229–238

Kujala M (1980) Runkopuun kuorellisen tilavuuskasvun laskentamenetelmä (Summary: A calculation method for measuring the volume growth over bark of stemwood). Folia Forestalia 441:1–8

Kuusela K (1972) Suomen metsävarat ja metsien omistus 1964–1970 sekä niiden kehittyminen 1920–70 (Forest resources and ownership in Finland 1964–70 and their development 1920–70). Communicationes Instituti Forestalis Fenniae 76(5):1–126

Kuusela K (1978) Suomen metsävarat ja metsien omistus 1971–1976 (Forest resources and ownership in Finland 1971–1976). Communicationes Instituti Forestalis Fenniae 93(6):1–170

Kuusela K, Salminen S (1991) Suomen metsävarat 1977–1984 ja niiden kehittyminen 1952–1980 (Forest resources of Finland in 1977–1984 and their development in 1952–1980). Acta Forestalia Fennica 220:1–84

Laki Lapin vajaatuottoisten metsien kunnostamisesta (1982) Laki 1057/1982. http://www.edilex.fi/saadokset/smur/19821057/muutos19961348. Accessed 20 Dec 2010 (In Finnish)

Lindgren M, Salemaa M, Tamminen P (1998) Metsien kunnon riippuvuus ympäristötekijöistä. In: Mälkönen E (ed) Ympäristömuutos ja metsien kunto. Metsien terveydentilan loppuraportti, vol 691. Metsäntutkimuslaitoksen tiedonantoja, Helsinki, 278 pp (In Finnish)

MELA ja metsälaskelmat (2010) Timber production potentials and MELA-system. http://www.metla.fi/metinfo/mela/index-en.htm. Accessed 7 Dec 2010 (In Finnish)

Metsätalouden kehittämiskeskus Tapio (1994) Luonnonläheinen metsänhoito. Metsänhoitosuositukset 1994. Metsäkeskus Tapion julkaisu 6/1994, 72 p

Metsätalouden kehittämiskeskus Tapio (2001) Hyvän metsänhoidon suositukset, 95 p (In Finnish)

National Land Survey of Finland (2003) Suomen pinta-ala kunnittain (The area of Finland by municipalities). National Land Survey of Finland, Helsinki, 1 Jan 2003

Nevalainen S (1999) Gremmeniella abietina in Finnish Pinus sylvestris Stands in 1986–1992: a study based on the National Forest Inventory. Scand J For Res 14(2):111–120

Nevalainen S (2002) The incidence of Gremmeniella abietina in relation to topography in southern Finland. Silva Fennica 36(2):459–473

Nevalainen S, Yli-Kojola H (2000) Extent of abiotic damage and its relation to defoliation of conifers in Finland. For Ecol Manage 135:229–235

Nuutinen T, Hirvelä H, Salminen O (2005) Alueelliset hakkuumahdollisuudet Suomessa. Metlan työraportteja/Working papers of the Finnish Forest Research Institute 13, 73 p. http://www.metla.fi/julkaisut/workingpapers/2005/mwp013.htm. Accessed 7 Dec 2010 (In Finnish)

Ojansuu R, Henttonen H (1983) Kuukauden keskilämpötilan, lämpösumman ja sademäärän paikallisten arvojen johtaminen Ilmatieteen laitoksen mittaustiedoista (Derivation of local values of monthly mean temperature, effective temperature sum, and precipitation, from the observations of the Finnish Meteorological Institute). Silva Fennica 17(2):143–150 (In Finnish)

Paavilainen E, Tiihonen P (1988) Suomen suometsät vuosina 1951–1984 (Summary: Peatland forests in Finland 1951–1984). Folia Forestalia 714:1–29

PEFC Finland (2010) PEFC standards in Finland. http://www.pefc.fi/pages/en/documents/standards.php. Accessed 10 Dec 2010

Penman J, Gytarsky M, Hiraishi T, Krug T, Kruger D, Pipatti R, Buendia L, Miwa K, Ngara T, Tanabe K, Wagner F (eds) (2003) Good practice guidance for land use, Land-use change and forestry. Intergovernmental Panel on Climate Change Working Group – National Greenhouse Gas Inventories Programme, Hayama, Japan

Siitonen J (2001) Forest management, coarse woody debris and saproxylic organisms: Fennoscandian boreal forests as an example. Ecol Bull 49:11–41

Siitonen M, Härkönen K, Hirvelä H, Jämsä J, Kilpeläinen H, Salminen O, Teuri M (1996) MELA handbook – 1996 edition. The Finnish Forest Institute, Research papers 622, Helinski, 452 p

Statistics Finland (2010) Greenhouse gas emissions in Finland 1990–2008. National inventory report to under the UNFCCC and the Kyoto Protocol. http://www.stat.fi/tup/khkinv/fin_nir_20100525.pdf. Accessed 1 Nov 2010

Tamminen P, Tomppo E (2008) Finnish forest soils. Working papers of the Finnish Forest Research Institute 100, Finland, 19 p. http://www.metla.fi/julkaisut/workingpapers/2008/mwp100.pdf. Accessed 20 December 2010

Tomppo E (1999) Forest resources of Finnish peatlands in 1951–1994. Int Peat J 9:38–44

Tomppo E (2000) Kasvupaikat ja puusto. In: Reinikainen A, Mäkipää R, Vanha-Majamaa I, Hotanen JP (eds.) Kasvit muuttuvassa metsäluonnossa. Kustannusosakeyhtiö Tammi, Helsinki, pp 60–83 (In Finnish)

References

Tomppo E (2005) Metsien puuvaranto ja puuston kasvu lisääntyneet Lapissa. In: Varmola M, Hyppönen M, Hokajärvi P (eds) Lapin metsien tulevaisuus. Metsäntutkimuslaitos, Metsäkeskus Lappi, Metsähallitus, Lapin Metsänhoitajat ry, Helsinki, pp 28–33 (In Finnish). ISBN 951-40-1966-0

Tomppo E (2006) Suomen suometsät 1951–2003. In: Ahti E, Kaunisto S, Moilanen M, Murtovaara I (eds) Suosta metsäksi. Suometsien ekologisesti ja taloudellisesti kestävä käyttö. Tutkimusohjelman loppuraportti. The Finnish Forest Research Institute, Research Papers 947, Helinski, pp 26–38 (In Finnish)

Tomppo E, Henttonen H (1996) Suomen metsävarat 1989–1994 ja niiden muutokset vuodesta 1951 lähtien. Metsätilastotiedote 354:1–18 (In Finnish)

Tomppo E, Henttonen H, Tuomainen T (2001) Valtakunnan metsien 8. inventoinnin menetelmä ja tulokset metsäkeskuksittain Pohjois-Suomessa 1992–94 sekä tulokset Etelä-Suomessa 1986–92 ja koko maassa 1986–94. Metsätieteen aikakauskirja 1B:99–248 (In Finnish)

Tomppo E, Gabler K, Schadauer K, Gschwantner T, Lanz A, Ståhl G, McRoberts R, Chirici G, Cienciala E, Winter S (2010) Summary of accomplishments. In: Tomppo E, Gschwantner T, Lawrence M, McRoberts RE (eds) National forest inventories: pathways for common reporting. Springer, Heidelberg, Dordrecht, London, New York, 672 p

Tuomainen T (2000) Männyn paksuuskasvun estimointi. Master thesis in forest management, University of Helsinki, Helsinki, 63 p

UNECE/FAO (2000) Forest resources of Europe, CIS, North America, Australia, Japan and New Zealand (industrialized temperate/boreal countries. UN-ECE/FAO contribution to the Global Forest Resources Assessment 2000. Main report, Timber and Forest Study Papers 17. United Nations, Geneva, 445 pp

Uotila A, Kankaanhuhta V (1999) Metsätuhojen tunnistus ja torjunta. Metsälehti kustannus, Helsinki, 215 p (In Finnish)

World Reference Base for Soil Resources (1998) World soil resources reports 84. FAO, ISRIC and ISSS, Rome

Yli-Kojola H, Nevalainen S (2006) Metsätuhojen esiintyminen Suomessa 1986–94. Metsätieteen aikakauskirja 1:97–180 (In Finnish)

Chapter 5
Discussion

5.1 The Development of Volume and Increment: New Estimates from NFI9

The 9th National Forest Inventory of Finland provided estimates concerning land areas, land use, forest resources and their structure, accomplished and required silvicultural, cutting regimes and forest health for practical forestry and forest industry decision making, as well as national for international statistics. It also provided estimates concerning a set of new characteristics. The inventory showed that the growing stock had continued its increase (Sect. 4.7). The annual increment of the growing stock had also increased from that in NFI8 (Sect. 4.8). The increase in the increment on mineral soils has mainly been the result of the changes in the density and structure of the forests (increases in the average increment). On the peatlands, drainage has also increased the forest area and improved the productivity of sites (Sect. 4.8.4). The intensified forestry regimes aimed at increasing both the increment and growing stock began in early 1960s.

The compatibility of NFI8 and NFI9 volume and increment estimates and drain statistics were checked using a concept called forest balance (Sect. 4.8.3). However, need was revealed for a further analysis of the components of the balance, e.g., drain statistics and/or increment estimates. The availability of the permanent plots in the entire country following the completion of NFI10 now makes it possible to analyse the accuracy of drain statistics.

The measurements and estimates of dead wood, including decayed dead wood, were included into NFI9 for the first time. Only pieces of ded wood with a length of at least of 1.3 m and with a diameter of at least 10 cm were included in the estimates (Sects. 2.20 and 4.13.3). The results showed considerable regional differences in the mean volume of the ded wood (Fig. 4.39) ranging from 1.2 m^3/ha (Etelä-Pohjanmaa) to 10.5 m³/ha (Lappi) (Fig. 2.1, Sect. 4.13.3, Fig. 4.38). The minimum dimensions of ded wood to be included in the estimates could be re-assessed from the ecological point of view. For example, ded wood shorter than 1.3 m can also be valuable for wood-inhabiting species (Siitonen 2001).

A group of characteristics describing key habitats was also new to NFI9 (Sects. 2.15 and 4.13.2). The most valuable habitats, from the ecological point of view, "habitats of particular significance", are protected by the Finnish Forest Act. The area of these Forest Act habitats corresponded to 1.7% of the area of the combined forest land, poorly productive forest land and unproductive land (Table A.41, Sect. 4.13). The area is somewhat larger than the area found in the inventories by Forestry Centres due to the reasons mentioned in Sects. 4.13.2 and 5.4.

5.2 Estimation and Error Estimation

Robust, design-based estimators were used for both the areas (Sect. 3.1) and for the parameters concerning growing stock and its increment (Sects. 3.2 and 3.3), as well as for the volume of dead wood (Sect. 3.4). The estimators do not require any distribution properties about the variables in question for being ratio-unbiased (ratios of two unbiased estimators). In the northernmost part of Finland, in sampling region 6 (Fig. 2.1b), the estimators had to be modified for the two-phase sampling employed. The error estimators (Sect. 3.5) are conservative (they do not underestimate the errors) under mild conditions. The error could be under-estimated in the case of periodic spatial variation in the variables, e.g., if the variation of land use classes or volume coincide with the spacing between locations of the sampling units. This can occur if sample plot clusters are located systematically on the valleys or tops of the ridges, and the tops and valleys differ with respect to the variables of interest. This problem has traditionally been addressed in the Finnish NFI when selecting the shape and orientation of the sampling units (earlier survey lines and current clusters).

Another principle in the estimation was that it should require very little ancillary data. The information for the area estimation, in addition to the field data, was the land area by municipalities, which was assumed to be error free, as well as ownership information for allocating the plots to the ownership categories (Cadastre 2010). Examples of data from other external sources were volume models (Sect. 3.2.3; Laasasenaho 1982) for the volume estimation, as well as models for the changes in stem form and bark thickness in increment estimation (Sect. 3.3.1; Kujala 1980). Neither map nor remote sensing data were used in, e.g., delineating land categories. The Finnish multi-source NFI utilises both these map and remote sensing data (Sect. 1.7). The estimation methods employed can thus be implemented and used in other circumstances in a quite straightforward way.

5.3 Comparisons of the NFI8 and NFI9 Designs

The new sampling design for NFI9 proved to be more efficient than the one used in NFI8. Despite the smaller number of sample plots, the sampling errors of the basic area and volume estimates were smaller than those in NFI8. The sampling errors of the forest area estimates for South Finland and North Finland were 472 and 679 km^2 while those in NFI8 were 510 and 683 km^2 (Table 5.1, Table A.1; Tomppo et al. 2001). The design was changed for NFI9 in South Finland and in the northernmost part of North Finland. In the greater part of North Finland, the design was already changed for NFI8, which explains the lower decrease in the errors from NFI8 to NFI9 compared to the decrease in South Finland. The number of the sample plots on forest land in the whole country in NFI9 was 57,457 and 60,248 in NFI8. The numbers on the combined forest land and poorly productive forest land were 62,266 and 65,932, respectively.

The sampling errors of the mean volume estimates were generally lower in NFI9 than in NFI8 (Table A.15a; Tomppo et al. 2001). Those for the total volume estimates, including the effects of the errors of the area estimates, were thus also lower in NFI9 than in NFI8 (Table 5.2, Table A.15a; Tomppo et al. 2001). In most cases, the

Table 5.1 The sampling errors (km^2) of the area estimates of land classes in the 8th and 9th National Forest Inventories

	Forest land	Poorly productive forest land	Un-productive land	Forest roads, depots, etc.	Total
South Finland					
NFI8	510	157	167	54	515
NFI9	472	127	145	50	479
North Finland					
NFI8	683	709	759	62	346
NFI9	679	566	596	67	296
Whole country					
NFI8	853	726	777	82	620
NFI9	827	580	614	83	563

Table 5.2 Relative standard error, i.e., coefficient of variation (%) of the total volume estimate by tree species groups on the combined forest land and poorly productive forest land in the 8th and 9th National Forest Inventories

	Coefficient of variation (%)					
	South Finland		North Finland		Whole country	
	NFI8	NFI9	NFI8	NFI9	NFI8	NFI9
Pine	1.00	0.86	1.70	1.34	0.90	0.75
Norway spruce	1.20	1.07	2.40	2.52	1.10	0.99
Silver birch	2.10	1.83	5.30	4.95	1.90	1.72
Downy birch	1.40	1.27	2.00	1.88	1.20	1.09
Other broadleaved	2.40	2.20	4.60	4.60	2.20	1.99
Total	0.70	0.63	1.30	0.99	0.60	0.53

coefficient of variation for total volume estimates in NFI9 decreased more than 10% compared to NFI8. Only the error of spruce volume in North Finland was slightly higher in NFI9 than in NFI8 (Table 5.2). Overall, the new design met the requirements and expectations in reducing sampling errors and measurement costs.

5.4 Some Experiences of the New Measurements

A set of new measurements in NFI9 concerned the assessment of biodiversity. The variables had been tested only in the trial inventories before the start of NFI9. The proper inventory revealed the need to modify some of the new measurements after the first and second field seasons. The sample plot radius for dead wood (Sect. 2.20) was initially 12.52 m. It was measured only on every second tally tree plot in the first year of NFI9 (1996). The search for lying ded wood, often in a state of advanced decayed, and checking the inclusion of standing ded wood was time consuming in such a big circle. The radius of the dead wood plot was therefore reduced to 7 m, and the plot was measured on every tally tree plot the centre point of which was on forest land or on poorly productive forest land. The total measured area was about 38% lower than in the method of the first year. Assessing the effect on the sampling error would need additional information, e.g., information about the within stand and between stand variation of ded wood.

Key habitat classes (Sect. 2.15) were checked, and revised if necessary, for each forest centre taking into account the assessed commonness of the habitat class in each region. In practice, the assessment of the commonness is difficult without a priori information. This concerns particularly the variable ecological value of the class and its Forest Act status. The conditions for that status are a regional rarity and for some key habitat classes also small size. In NFI9, however, the commonness of a certain habitat class or habitat characteristics was not taken into account, except for eutrophic fens that were not given Forest Act status in Lapland. Only in 1999, the small size requirement was included in the NFI9 field instructions. Previously, also large sites of herb-rich forests and oligo- and ombrotrophic mires with only sparse tree stands may have been regarded as Forest Act habitats, more precisely classes I, B, V, C, and E in Table 2.23.

5.5 Experiences with the New Measurement Devices

Some new measurement devices were also introduced in NFI9 (Sect. 2.21). The height measurement instrument (Suunto Hypsometer) was replaced by a Haglöf Vertex-gauge, Vertex III-60 in 2001. The new device presumes both personal calibration and daily calibration to account for temperature variations. The height, and particularly height increment measurements turned out to be somewhat high after the first field season for many field crews. A more thorough check of the height increments resulted in control measurements in autumn 2002. The height and height increments of the sub-sample trees were re-measured. The details are presented in Sect. 2.22. The height and height correction models were estimated using the measured data. The models were applied to the rest of the sample trees, i.e., those not re-measured (Sects. 2.22.1 and 2.22.2; Tomppo et al. 2003).

Since 2001, the measurements of the tree rings for diameter increment of 5 years and the age at breast height (partially) were changed to a semi-automatic image analysis system. A majority of the increment cores of the sample trees were measured using a WinDENDRO ring scanner and software version 6.2 (Sect. 2.21, Windendro 2010). The hardware and software were thoroughly tested in advance (Tuomainen 2000). The measurements were precise and the system facilitated the work.

The first attempt to locate the field plots with GPS (Global Positioning System) devices was carried out in the first year of NFI9 (1996). Use of the Garmin GPS 45 single channel device was too time consuming and the locations of the plots not precise enough and so it was not used (Sect. 2.3). The field plots in South Finland (1996–2000) were located using map measurements. When NFI9 proceeded to North Finland (2001), a sufficiently fast and precise GPS receiver (Trimble ACE II) was available. The GPS system was assembled at Metla installing the receiver into a backpack together with battery and cabling. A field computer was used to display and store the GPS data. An external antenna was mounted on the backpack (Sect. 2.3).

5.6 Changes in the Design for NFI10

An important condition for a cost-efficient design is that a field plot cluster is on average a single day's workload. For many crews, the clusters in sampling region 3 required more than one day. Therefore, the number of the plots per cluster was slightly reduced in regions 3 for NFI10 (2004–2008). The number of the plots per cluster was also reduced in region 2, in order to reduce the total number of the plots so that the whole inventory could be carried out in five years.

A more fundamental change in NFI10 concerned the way the fieldwork proceeded. Since the fifth inventory (1964–1970), the inventory had proceeded by regions. The procedure for NFI0 was changed to a system in which the entire country was covered by the field measurements each year. The main reason was the need to compute basic forest resource estimates annually for the entire country in such a way that at least a part of the measurements come from the year in question (Tomppo 2009). In the new design, one fifth of the clusters were measured annually. Each fifth fraction should be, in principle, a systematic sample of the clusters over the entire country. This design is sometimes called a panel system (Tomppo et al. 2010). In the new system, the inventory rotation was shortened from almost 10 to 5 years. The need to shift the locations of the clusters with temporary plots caused some changes. Some exceptions from a pure panel system had to be made in NFI10: all of the clusters in sampling region 1 (Fig. 2.2) were measured in one field season, in 2007. The plots in region 6 were not measured in NFI10; the data from NFI9 were used. The changes in the growing stock and land use are slow in the northernmost Finland. The first year of NFI10, 2004, was exceptional because of the budget constraints. The plots were measured only in sampling density regions 2 and 3 in the southern part of the country.

NFI11 (2009–2013) is proceeding in a principle in a same a way as NFI10. A need for precise and up-to-date estimates about the land use changes, often small in area, may cause a further revision the sampling design, and possibly the use of additional information in estimation, not only in multi-source inventory but also in the inventory based on the field data.

References

Kujala M (1980) Runkopuun kuorellisen tilavuuskasvun laskentamenetelmä (Summary: a calculation method for measuring the volume growth over bark of stemwood). Folia Forestalia 441:1–8

Cadastre (2010) National Land Survey of Finland. http://www.maanmittauslaitos.fi/digituotteet/kiinteistorekisteri. Accessed 15 Dec 2010 (In Finnish)

Laasasenaho J (1982) Seloste: Männyn, kuusen ja koivun runkokäyrä – ja tilavuusyhtälöt (Taper curve and volume functions for pine, spruce and birch). Communicationes Instituti Forestalis Fenniae 108:1–74

Siitonen J (2001) Forest management, coarse woody debris and saproxylic organisms: Fennoscandian boreal forests as an example. Ecol Bull 49:11–41

Statistics Finland (2005). Statistical Yearbook of Finland 2005. http://pxweb2.stat.fi/sahkoiset_julkaisut/vuosikirja2005/alku.htm. Accessed 1 Nov 2010

Tomppo E (2009) The Finnish National Forest Inventory. In: McRoberts RE, Reams GA, Van Deusen PA, McWilliams WH (eds.) Proceedings of the eighth annual forest inventory and analysis symposium, Monterey, CA, 16–19 Oct 2006. USDA Forest Service General Technical Report WO-79, pp 39–46

Tomppo E, Henttonen H, Tuomainen T (2001) Valtakunnan metsien 8. inventoinnin menetelmä ja tulokset metsäkeskuksittain Pohjois-Suomessa 1992–94 sekä tulokset Etelä-Suomessa 1986–92 ja koko maassa 1986–94. Metsätieteen aikakauskirja 1B/2001:99–248 (In Finnish)

Tomppo E, Tuomainen T, Henttonen H, Ihalainen A, Tonteri T (2003) Kainuun metsäkeskuksen alueen metsävarat 1969-2001. Metsätieteen aikakauskirja 2B:169–256 (In Finnish)

Tomppo E, Gschwantner T, McRoberts RE, Lawrence M (2010) National forest inventories – pathways for common reporting. Springer, Heidelberg. ISBN 978-90-481-3232-4

Tuomainen, T (2000) Männyn paksuuskasvun estimointi. Master thesis in forest management, University of Helsinki, 63 p

Windendro (2010) http://www.regentinstruments.com/products/dendro/DENDRO.html. Accessed 10 Nov 2010

Appendix

Note 1. Notation "." in a cell means an item is illogical or impossible on the basis of inventory instructions (e.g. site fertlity classes 1-6 on poorly productive forest land on mineral soil).

Note 2. Notation "–" in a cell means not possible to estimate, e.g. the mean diameter for a forest and tree category without any tally tree in the sample.

Note 3. Notation "s.e." means sampling error.

Note 4. The estimate for tree species Pine includes also the other coniferous species than Norway Spruce.

Table A.1 The areas of land-use classes according to the national definitions and areas of FAO land classes

	National classification						Total land area	FAO classification		
	Forest land	Poorly productive forest land	Unproductive land	Forest roads, depots, etc.	Forestry land total	Other land		Forest	Other wooded land	Other land
South Finland										
Area, km^2	111,671	4,280	3,526	858	120,334	34,279	154,613	115,310	1,094	38,209
Sampling error, km^2	472	127	145	50	479	479		479	59	481
Land class of total land area, %	72.2	2.8	2.3	0.6	77.8	22.2	100.0	74.6	0.7	24.7
North Finland										
Area, km^2	91,707	22,422	28,030	677	142,836	7,024	149,861	109,559	7,156	33,146
Sampling error, km^2	679	566	596	67	296	296		651	348	645
Land class of total land area, %	61.2	15.0	18.7	0.5	95.3	4.7	100.0	73.1	4.8	22.1
Whole country										
Area, km^2	203,378	26,701	31,556	1,536	263,171	41,303	304,474	224,869	8,250	71,355
Sampling error, km^2	827	580	614	83	563	563		808	353	804
Land class of total land area, %	66.8	8.8	10.4	0.5	86.4	13.6	100.0	73.9	2.7	23.4

Appendix

Table A.2 The areas of land-use class conversions during the previous 10-year period

Present land-use class	Land-use class 10 years ago					Other land-use classes			Total area of land-use class
	Forestry land						Of which now reverting to forestry land	Water	
	Forest land	Poorly productive forest land	Unproductive land	Forest roads, depots, etc.					
	km²								
South Finland									
Forest land	110,670	182	11	3		804	.	0	111,671
Poorly productive forest land	68	4,143	69	0		0	.	0	4,280
Unproductive land	0	20	3,503	0		0	.	3	3,526
Forest roads, depots, etc.	223	3	0	624		9	.	0	858
Other land-use classes	1,056	60	31	3		32,397	731	0	34,279
North Finland									
Forest land	91,037	456	18	14		183	.	0	91,707
Poorly productive forest land	170	22,091	143	0		18	.	0	22,422
Unproductive land	0	92	27,934	0		0	.	4	28,030
Forest roads, depots, etc.	134	0	4	539		0	.	0	677
Other land-use classes	155	67	60	0		6,257	485	0	7,024
Whole country									
Forest land	201,707	638	29	17		987	.	0	203,378
Poorly productive forest land	239	26,233	212	0		18	.	0	26,701
Unproductive land	0	112	31,437	0		0	.	6	31,556
Forest roads, depots, etc.	358	3	4	1,163		9	.	0	1,536
Other land-use classes	1,212	127	91	3		38,654	1,216	0	41,303

Table A.3 The area of forestry land by ownership categories

	Private Area, km²	s.e., km²	Proportion of land class area, %	Communities Area, km²	s.e., km²	Proportion of land class area, %	Companies Area, km²	s.e., km²	Proportion of land class area, %	State Area, km²	s.e., km²	Proportion of land class area, %	Total Area, km²	s.e., km²
South Finland														
Forest land	83,379	585	74.7	6,609	260	5.9	13,397	411	12.0	8,287	302	7.4	111,671	472
Poorly productive forest land	2,763	101	64.6	359	38	8.4	522	47	12.2	635	55	14.8	4,280	127
Unproductive land	1,640	79	46.5	573	54	16.2	334	44	9.5	979	96	27.8	3,526	145
Forest roads, depots, etc.	518	38	60.4	65	15	7.6	116	18	13.5	159	25	18.5	858	50
Total	88,300	605	73.4	7,606	282	6.3	14,369	431	11.9	10,059	346	8.4	120,334	479
North Finland														
Forest land	39,015	746	42.5	4,120	371	4.5	4,758	270	5.2	43,814	840	47.8	91,707	679
Poorly productive forest land	5,894	244	26.3	769	114	3.4	691	80	3.1	15,068	533	67.2	22,422	566
Unproductive land	4,505	224	16.1	753	97	2.7	582	76	2.1	22,190	584	79.2	28,030	596
Forest roads, depots, etc.	276	41	40.7	57	18	8.5	42	12	6.2	302	47	44.6	677	67
Total	49,691	913	34.8	5,699	469	4.0	6,074	327	4.3	81,373	988	57.0	142,836	296
Whole country														
Forest land	122,394	948	60.2	10,729	453	5.3	18,155	492	8.9	52,100	893	25.6	203,378	827
Poorly productive forest land	8,657	264	32.4	1,128	121	4.2	1,213	93	4.5	15,703	536	58.8	26,701	580
Unproductive land	6,145	238	19.5	1,326	111	4.2	916	88	2.9	23,169	592	73.4	31,556	614
Forest roads, depots, etc.	794	56	51.7	123	24	8.0	158	22	10.3	461	53	30.0	1,536	83
Total	137,991	1,095	52.4	13,305	548	5.1	20,442	541	7.8	91,433	1,047	34.7	263,171	563

Table A.4 The areas of the activities which affect forest management on forestry land

	Forest land			Poorly productive forest land			Unproductive land			Forest roads, depots, etc.			Total		
	Area, km²	s.e., km²	Proportion of land class area, %	Area, km²	s.e., km²	Proportion of land class area, %	Area, km²	s.e., km²	Proportion of land class area, %	Area, km²	s.e., km²	Proportion of land class area, %	Area, km²	s.e., km²	Proportion of land class area, %
South Finland															
Nature conservation	3,030	174	2.7	401	46	9.4	992	103	28.1	12	7	1.3	4,435	240	3.7
Outdoor recreation	866	108	0.8	22	8	0.5	14	9	0.4	0	.	0.0	903	113	0.8
Military use	399	66	0.4	20	9	0.5	3	3	0.1	8	5	1.0	430	70	0.4
Land-use planning	3,127	158	2.8	83	15	1.9	105	19	3.0	17	7	1.9	3,332	163	2.8
Other[a]	94	28	0.1	9	5	0.2	6	3	0.2	11	9	1.3	120	32	0.1
Total	7,517	260	6.7	535	50	12.5	1,120	105	31.8	48	14	5.5	9,219	310	7.7
North Finland															
Nature conservation	10,365	575	11.3	6,953	490	31.0	14,501	714	51.7	4	4	0.5	31,823	1,136	22.3
Outdoor recreation	77	35	0.1	18	9	0.1	21	9	0.1	0	.	0.0	116	50	0.1
Military use	721	130	0.8	50	17	0.2	14	11	0.1	4	4	0.5	789	146	0.6
Land-use planning	842	119	0.9	32	12	0.1	11	8	0.0	4	4	0.5	888	123	0.6
Other[a]	1,105	203	1.2	530	128	2.4	290	124	1.0	18	18	2.7	1,942	346	1.4
Total	13,110	613	14.3	7,582	490	33.8	14,837	703	52.9	29	19	4.2	35,558	1,133	24.9
Whole country															
Nature conservation	13,396	601	6.6	7,354	492	27.5	15,493	721	49.1	15	8	1.0	36,258	1,161	13.8
Outdoor recreation	944	114	0.5	40	12	0.1	35	13	0.1	0	.	0.0	1,019	123	0.4
Military use	1,120	146	0.6	69	19	0.3	17	12	0.1	12	6	0.8	1,218	162	0.5
Land-use planning	3,969	198	2.0	115	19	0.4	116	21	0.4	20	8	1.3	4,220	204	1.6
Other[a]	1,199	205	0.6	538	128	2.0	295	124	0.9	29	20	1.9	2,062	348	0.8
Total	20,627	666	10.1	8,116	493	30.4	15,957	711	50.6	76	24	5.0	44,777	1,175	17.0

[a]Other: monuments of antiquity; areas reserved for tree breeding and research; undefined

Table A.5 The areas of principal site classes and site fertility classes on forest land, on poorly productive forest land and on unproductive land

Site fertility class	1 Area km²	1 s.e. km²	1 %[a]	2 Area km²	2 s.e. km²	2 %[a]	3 Area km²	3 s.e. km²	3 %[a]	4 Area km²	4 s.e. km²	4 %[a]	5 Area km²	5 s.e. km²	5 %[a]	6 Area km²	6 s.e. km²	6 %[a]	7 Area km²	7 s.e. km²	7 %[a]	8 Area km²	8 s.e. km²	8 %[a]	Total Area km²	Total s.e. km²
South Finland																										
Forest land																										
Mineral soil	3,392	112	4.0	21,066	288	24.8	41,006	384	48.3	16,624	277	19.6	1,537	87	1.8	63	22	0.1	1,286	68	1.5	.	.	.	84,974	459
Spruce mires	583	41	4.9	3,478	108	29.5	6,984	156	59.2	753	50	6.4	.	.	.	85	.	.	.	.	.	.	.	.	11,798	203
Pine mires	23	8	0.2	244	28	1.6	1,719	80	11.5	7,941	183	53.3	4,887	147	32.8	.	18	0.6	.	.	.	.	.	.	14,899	271
Poorly productive forest land																										
Mineral soil	.	.	.	.	.	.	.	.	.	.	.	.	.	.	.	.	.	.	1,599	75	100.0	0	.	0.0	1,599	75
Spruce mires	17	7	11.3	99	17	67.3	26	9	17.5	6	4	3.8	.	.	.	.	.	.	.	.	.	.	.	.	147	22
Pine mires	20	7	0.8	32	9	1.2	124	18	4.9	245	27	9.6	1,533	74	60.5	581	48	22.9	.	.	.	.	.	.	2,534	97
Unproductive land																										
Mineral soils	.	.	.	.	.	.	.	.	.	.	.	.	.	.	.	.	.	.	709	47	100.0	0	.	0.0	709	47
Spruce mires	3	3	6.8	30	10	72.7	8	5	20.5	0	.	0.0	.	.	.	.	.	.	.	.	.	.	.	.	41	11
Pine mires	6	6	0.9	17	12	2.6	40	15	6.3	42	12	6.7	226	28	35.7	303	35	47.8	.	.	.	.	.	.	635	51
Treeless peatland	51	15	2.4	408	42	19.1	290	32	13.5	280	42	13.1	711	64	33.2	402	47	18.8	.	.	.	.	.	.	2,141	120
North Finland																										
Forest land																										
Mineral soil	466	57	0.7	1,744	109	2.6	35,020	617	52.3	26,084	613	39.0	3,278	249	4.9	18	12	0.0	312	66	0.5	.	.	.	66,922	705
Spruce mires	542	58	6.3	3,059	151	35.4	4,589	165	53.1	459	52	5.3	.	.	.	11	6	.	.	.	.	.	.	.	8,649	248
Pine mires	231	48	1.4	743	76	4.6	2,356	120	14.6	10,485	258	65.0	2,309	107	14.3	.	.	0.1	.	.	.	.	.	.	16,136	331
Poorly productive forest land																										
Mineral soil	.	.	.	.	.	.	.	.	.	.	.	.	.	.	.	.	.	.	1,868	261	25.4	5,484	439	74.6	7,352	458
Spruce mires	342	55	14.0	836	86	34.2	589	67	24.1	677	75	27.7	.	.	.	.	.	.	.	.	.	.	.	.	2,445	149
Pine mires	564	73	4.5	770	82	6.1	2,023	117	16.0	4,935	193	39.1	3,896	171	30.9	438	47	3.5	.	.	.	.	.	.	12,625	328

Unproductive land																			
Mineral soil	206	66	33.0	250	60	40.2	103	38	16.5	.	64	21	10.3	.	.	.	.	.	1,114	271	11.8	8,344	530	88.2	9,458	531
Spruce mires	.	.	.	.	.	.	.	.	.	.	.	.	.	.	.	.	.	.	.	.	.	.	.	.	623	113
Pine mires	146	39	3.8	288	54	7.5	1,297	120	33.6	765	1,094	112	28.3	84	19.8	270	49	7.0	.	.	.	.	.	.	3,860	212
Treeless peatland	745	136	5.3	2,107	179	15.0	6,592	300	46.8	1,684	2,680	198	19.0	130	12.0	281	41	2.0	.	.	.	.	.	.	14,089	476
Whole country																										
Forest land																										
Mineral soil	3,858	126	2.5	22,811	308	15.0	76,025	727	50.1	4,815	42,708	673	28.1	264	3.2	81	25	0.1	1,598	95	1.1	.	.	.	151,896	841
Spruce mires	1,125	71	5.5	6,537	185	32.0	11,573	226	56.6	.	1,213	72	5.9	.	.	.	.	.	.	.	.	.	.	.	20,447	320
Pine mires	254	49	0.8	987	81	3.2	4,075	144	13.1	7,196	18,426	316	59.4	182	23.2	95	19	0.3	.	.	.	.	.	.	31,035	428
Poorly productive forest land																										
Mineral soil	.	.	.	.	.	.	.	.	.	.	.	.	.	.	.	.	.	.	3,467	272	38.7	5,484	439	61.3	8,951	464
Spruce mires	359	56	13.9	935	87	36.1	615	67	23.7	.	683	75	26.4	.	.	.	.	.	.	.	.	.	.	.	2,592	151
Pine mires	584	74	3.9	802	83	5.3	2,147	119	14.2	5,429	5,179	195	34.2	186	35.8	1,018	67	6.7	.	.	.	.	.	.	15,159	342
Unproductive land																										
Mineral soil	208	66	31.4	280	61	42.2	111	38	16.8	.	64	21	9.7	.	.	.	.	.	1,824	275	17.9	8,344	530	82.1	10,167	533
Spruce mires	.	.	.	.	.	.	.	.	.	.	.	.	.	.	.	.	.	.	.	.	.	.	.	.	664	113
Pine mires	152	39	3.4	305	55	6.8	1,337	121	29.7	992	1,136	112	25.3	88	22.1	573	60	12.7	.	.	.	.	.	.	4,495	218
Treeless peatland	796	137	4.9	2,515	184	15.5	6,882	302	42.4	2,395	2,960	202	18.2	145	14.8	683	62	4.2	.	.	.	.	.	.	16,230	491
Total																										
Mineral soil	3,858	126	2.3	22,811	308	13.3	76,025	727	44.5	4,815	42,708	673	25.0	264	2.8	81	25	0.0	6,889	448	4.0	13,828	594	8.1	171,014	870
Spruce mires	1,693	115	7.1	7,751	219	32.7	12,299	243	51.9	.	1,960	108	8.3	.	.	.	.	.	.	.	.	.	.	.	23,703	383
Pine mires	990	101	2.0	2,094	130	4.1	7,559	230	14.9	13,617	24,742	387	48.8	289	26.9	1,687	93	3.3	.	.	.	.	.	.	50,688	598
Treeless peatland	796	137	4.9	2,515	184	15.5	6,882	302	42.4	2,395	2,960	202	18.2	145	14.8	683	62	4.2	.	.	.	.	.	.	16,230	491

Site fertility class: 1 Herb rich sites on mineral soils and eutrophic mires, *2* Herb rich heath sites on mineral soils and mesotrophic mires, *3* Mesic forests on mineral soils and meso-oligotrophic eutrophic mires, *4* Sub-xeric sites on mineral soils and oligotrophic eutrophic mires, *5* Xeric sites on mineral soils and oligo-ombrotrophic mires, *6* Barren sites on mineral soils and *Sphagnum fuscum* dominated (ombrotrophic) mires, *7* Rocky and sandy mineral soils and alluvial lands, *8* Summit and field forest

[a] Proportion of the principal site class area

Table A.6 The areas of soil types by site fertility classes on mineral soil on forest land, and on poorly productive forest land and unproductive land

			Forest land									Poorly productive forest land and unproductive land	
			Site fertility class							Total			
	Soil type	1	2	3	4	5	6	7			%	km²	%
		km²											
South Finland	Organic	162	204	210	121	3	0	3	703		0.8	18	0.8
	Bedrock	17	103	680	955	207	21	1,075	3,057		3.6	2,017	87.4
	Stony soil	6	80	204	144	20	0	39	492		0.6	97	4.2
	Till total	1,453	14,525	33,308	11,276	648	6	143	61,358		72.2	56	2.4
	Fine till (<0.06 mm)	370	2,740	2,155	237	3	0	0	5,504		6.5	3	0.1
	Medium coarse till (0.06–0.6 mm)	1,062	11,626	30,274	10,146	535	6	138	53,787		63.3	36	1.6
	Coarse till (>0.6 mm)	21	159	880	893	109	0	6	2,068		2.4	17	0.7
	Sorted soil total	1,754	6,155	6,604	4,128	659	37	26	19,363		22.8	120	5.2
	Fine sorted soil (<0.06 mm)	1,366	3,779	1,668	181	9	0	3	7,006		8.2	70	3.0
	Medium coarse sorted soil (0.06–0.6 mm)	382	2,307	4,671	3,623	526	25	21	11,555		13.6	46	2.0
	Coarse sorted soil (>0.6 mm)	6	69	266	323	124	12	3	802		0.9	4	0.2
	Site fertility class total	3,392	21,066	41,006	16,624	1,537	63	1,286	84,974		100.0	2,308	100.0
North Finland	Organic	28	21	212	127	11	0	4	403		0.6	56	0.3
	Bedrock	0	4	42	174	14	0	121	355		0.5	360	2.1
	Stony soil	0	0	138	317	7	0	98	560		0.8	2,385	14.2
	Till total	159	1,218	30,221	20,836	1,930	0	72	54,436		81.3	12,550	74.7
	Fine till (<0.06 mm)	43	202	2,847	569	11	0	0	3,670		5.5	1,146	6.8
	Medium coarse till (0.06–0.6 mm)	116	1,002	27,185	19,820	1,812	0	72	50,007		74.7	10,888	64.8
	Coarse till (>0.6 mm)	0	14	190	447	107	0	0	758		1.1	516	3.1

Appendix

		1	2	3	4	5	6	7	8	Total		a
	Sorted soil total	279	501	4,406	4,630	1,316	18	18	18	11,167	16.7	1,459	8.7
	Fine sorted soil (<0.06 mm)	177	257	969	246	11	0	0	0	1,659	2.5	493	2.9
	Medium coarse sorted soil (0.06–0.6 mm)	102	244	3,310	4,142	1,166	18	11	11	8,992	13.4	958	5.7
	Coarse sorted soil (>0.6 mm)	0	0	127	241	140	0	7	7	516	0.8	7	0.0
	Site fertility class total	466	1,744	35,020	26,084	3,278	18	312	312	66,922	100.0	16,810	100.0
Whole country	Organic	191	225	422	248	13	0	6	6	1,106	0.7	74	0.4
	Bedrock	17	106	722	1,129	221	21	1,196	1,196	3,413	2.2	2,377	12.4
	Stony soil	6	80	342	461	27	0	137	137	1,053	0.7	2,483	13.0
	Till total	1,612	15,743	63,529	32,112	2,578	6	215	215	115,794	76.2	12,606	65.9
	Fine till (<0.06 mm)	412	2,942	5,001	805	13	0	0	0	9,174	6.0	1,148	6.0
	Medium coarse till (0.06–0.6 mm)	1,178	12,628	57,459	29,967	2,348	6	209	209	103,794	68.3	10,925	57.1
	Coarse till (>0.6 mm)	21	173	1,069	1,340	216	0	6	6	2,826	1.9	533	2.8
	Sorted soil total	2,033	6,656	11,010	8,757	1,976	54	44	44	30,531	20.1	1,579	8.3
	Fine sorted soil (<0.06 mm)	1,543	4,037	2,636	428	19	0	3	3	8,666	5.7	563	2.9
	Medium coarse sorted soil	484	2,551	7,981	7,765	1,692	43	31	31	20,547	13.5	1,004	5.3
	Coarse sorted soil (>0.6 mm)	6	69	393	564	265	12	10	10	1,318	0.9	11	0.1
	Site fertility class total	3,858	22,811	76,025	42,708	4,815	81	1,598	1,598	151,896	100.0	19,118	100.0

Site fertility class *1* Herb rich sites on mineral soils and eutrophic mires, *2* Herb rich heath sites on mineral soils and mesotrophic mires, *3* Mesic forests on mineral soils and meso-oligotrophic eutrophic mires, *4* Sub-xeric sites on mineral soils and oligotrophic eutrophic mires, *5* Xeric sites on mineral soils and oligo-ombrotrophic mires, *6* Barren sites on mineral soils and *Sphagnum fuscum* dominated (ombrotrophic) mires, *7* Rocky and sandy mineral soils and alluvial lands, *8* Summit and field forest

[a]Proportion of the principal site class area

Table A.7 The thickness of peat layer on peatland

Thickness of peat, cm	Principal site class									Total		
	Spruce mires			Pine mires			Treeless peatland					
	Area, km²	Proportion, %	Mean thickness, cm	Area, km²	Proportion, %	Mean thickness, cm	Area, km²	Proportion, %	Mean thickness, cm	Area, km²	Proportion, %	Mean thickness, cm
South Finland												
–30	5,011	41.8	17	2,685	14.9	19	58	2.7	15	7,754	24.1	18
31–50	2,364	19.7	42	2,122	11.7	43	73	3.4	44	4,560	14.2	42
51–100	2,315	19.3	77	3,632	20.1	79	277	12.9	84	6,223	19.3	79
101–200	1,508	12.6	147	4,653	25.8	154	474	22.1	158	6,635	20.6	153
201–300	415	3.5	248	2,640	14.6	254	394	18.4	261	3,449	10.7	254
301–399	152	1.3	349	978	5.4	345	182	8.5	346	1,312	4.1	345
Over 400	220	1.8	–	1,358	7.5	–	684	31.9	–	2,261	7.0	–
North Finland												
–30	5,625	48.0	17	7,832	24.0	18	1,311	9.3	15	14,768	25.3	18
31–50	2,258	19.3	42	5,810	17.8	43	1,331	9.4	44	9,399	16.1	43
51–100	2,104	18.0	76	8,497	26.0	78	3,194	22.7	81	13,794	23.6	78
101–200	1,203	10.3	145	6,927	21.2	148	4,451	31.6	153	12,581	21.5	149
201–300	340	2.9	257	2,236	6.9	252	2,236	15.9	255	4,812	8.2	254
301–399	92	0.8	351	777	2.4	345	827	5.9	347	1,696	2.9	346
Over 400	95	0.8	–	541	1.7	–	739	5.2	–	1,375	2.4	–
Whole country												
–30	10,636	44.9	17	10,518	20.7	18	1,369	8.4	15	22,522	24.9	18
31–50	4,622	19.5	42	7,932	15.6	43	1,404	8.7	44	13,959	15.4	43
51–100	4,419	18.6	77	12,129	23.9	78	3,470	21.4	82	20,018	22.1	78
101–200	2,711	11.4	146	11,580	22.8	151	4,925	30.3	154	19,216	21.2	151
201–300	755	3.2	251	4,876	9.6	254	2,630	16.2	256	8,262	9.1	254
301–399	245	1.0	349	1,754	3.5	345	1,009	6.2	347	3,008	3.3	346
Over 400	315	1.3	–	1,899	3.7	–	1,422	8.8	–	3,636	4.0	–

Table A.8 The areas of principal site classes by drainage situation on forest land, on poorly productive forest land and on unproductive land

	Land class																		
	Forest land					Poorly productive forest land					Unproductive land					Total				
			Proportion of					Proportion of					Proportion of					Proportion of	
			Principal site class	Land class				Principal site class	Land class				Principal site class	Land class				Principal site class	Land class
Principal site class Drainage stage	Area, km²	s.e. km²	area %	area	Area km²	s.e. km²	area %	area	Area km²	s.e. km²	area %	area	Area km²	s.e. km²	area %	area
South Finland																				
Mineral soil																				
Undrained	76,549	450	90	69	1,599	75	100	37	698	46	98	20	78,846	456	90	66
Drained	8,425	175	10	8	0	.	0	0	11	7	2	0	8,436	175	10	7
Total	84,974	459	100	76	1,599	75	100	37	709	47	100	20	87,282	464	100	73
Peatland																				
Undrained	3,263	103	12	3	1,310	69	49	31	2,584	134	92	73	7,157	197	22	6
Drained	23,433	322	88	21	1,371	70	51	32	233	26	8	7	25,038	343	78	21
Recently drained	826	52	3	1	441	36	16	10	156	21	6	4	1,423	68	4	1
Transforming	12,192	237	46	11	887	56	33	21	77	15	3	2	13,157	255	41	11
Transformed	10,415	203	39	9	43	11	2	1	0	.	0	0	10,458	203	32	9
Total	26,697	328	100	24	2,681	100	100	63	2,817	138	100	80	32,194	383	100	27
Total	111,671	472	.	100	4,280	127	.	100	3,526	145	.	100	119,476	477	.	100
North Finland																				
Mineral soil																				
Undrained	62,065	710	93	68	7,349	458	100	33	9,458	531	100	34	78,872	742	94	55
Drained	4,857	179	7	5	4	4	0	0	0	.	0	0	4,860	179	6	3
Total	66,922	705	100	73	7,352	458	100	33	9,458	531	100	34	83,732	736	100	59

(continued)

Table A.8 (continued)

Principal site class Drainage stage	Forest land Area, km²	s.e. km²	Proportion of Principal site class area %	Proportion of Land class area	Poorly productive forest land Area km²	s.e. km²	Proportion of Principal site class area %	Proportion of Land class area	Unproductive land Area km²	s.e. km²	Proportion of Principal site class area %	Proportion of Land class area	Total Area km²	s.e. km²	Proportion of Principal site class area %	Proportion of Land class area
Peatland																
Undrained	5,197	191	21	6	11,221	323	74	50	17,850	554	96	64	34,268	693	59	24
Drained	19,588	376	79	21	3,848	180	26	17	722	73	4	3	24,159	440	41	17
Recently drained	445	55	2	0	521	56	3	2	297	55	2	1	1,263	101	2	1
Transforming	14,143	331	57	15	3,289	162	22	15	425	50	2	2	17,856	394	31	13
Transformed	5,001	171	20	5	39	18	0	0	0	.	0	0	5,040	172	9	4
Total	24,785	399	100	27	15,070	360	100	67	18,572	557	100	66	58,427	730	100	41
Total	91,707	679		100	22,422	566		100	28,030	596	.	100	142,159	298		100
Whole country																
Mineral soil																
Undrained	138,614	841	91	68	8,947	464	100	34	10,156	533	100	32	157,718	871	92	60
Drained	13,281	250	9	7	4	4	0	0	11	7	0	0	13,296	250	8	5
Total	151,896	841	100	75	8,951	464	100	34	10,167	533	100	32	171,014	870	100	65
Peatland																
Undrained	8,460	217	16	4	12,531	330	71	47	20,433	570	96	65	41,424	720	46	16
Drained	43,022	495	84	21	5,220	193	29	20	955	78	4	3	49,197	558	54	19
Recently drained	1,271	75	2	1	961	67	5	4	453	59	2	1	2,686	122	3	1
Transforming	26,335	407	51	13	4,176	172	24	16	502	52	2	2	31,013	469	34	12
Transformed	15,416	266	30	8	82	21	0	0	0	.	0	0	15,498	266	17	6
Total	51,482	516	100	25	17,751	374	100	66	21,389	574	100	68	90,621	824	100	35
Total	203,378	827		100	26,701	580		100	31,556	614	.	100	261,635	563		100

Appendix

Table A.9 The areas of dominant tree species on forest land and on poorly productive forest land in whole country

Dominant tree species	Forest land Area, km²	s.e. km²	Proportion of land class area %	Poorly productive forest land Area, km²	s.e. km²	Proportion of land class area %	Total Area, km²	s.e. km²	Proportion of land class area %
Treeless	2,631	108	1.3	25	10	0.1	2,656	109	1.2
Scots pine	132,951	885	65.4	18,632	440	69.8	151,584	945	65.9
Norway spruce	48,290	555	23.7	2,407	194	9.0	50,698	614	22.0
Silver birch	5,165	133	2.5	22	7	0.1	5,187	134	2.3
Downy birch	12,628	273	6.2	5,506	395	20.6	18,134	476	7.9
European aspen	659	47	0.3	3	2	0.0	662	47	0.3
Grey alder	587	45	0.3	35	13	0.1	622	48	0.3
Black alder	142	18	0.1	45	11	0.2	187	21	0.1
European mountain ash	25	9	0.0	0	.	0.0	25	9	0.0
Goat willow	27	9	0.0	25	12	0.1	52	15	0.0
Lodgepole pine	88	17	0.0	0	.	0.0	88	17	0.0
Swiss stone pine	4	4	0.0	0	.	0.0	4	4	0.0
Larch	149	25	0.1	0	.	0.0	149	25	0.1
Other coniferous	3	3	0.0	0	.	0.0	3	3	0.0
European ash	8	4	0.0	0	.	0.0	8	4	0.0
Pedunculate oak	8	6	0.0	1	1	0.0	10	6	0.0
Norway maple	3	3	0.0	0	.	0.0	3	3	0.0
Other broadleaved	9	5	0.0	0	.	0.0	9	5	0.0
Total	203,378	827	100.0	26,701	580	100.0	230,079	802	100.0

Table A.10 The areas of dominant tree species on forest land by the proportion of the dominant tree species

Proportion of the dominant tree species

Dominant tree species	Over 95% km²	%	75–95% km²	%	Under 75% km²	%	Total km²	%
South Finland								
Pine	28,813	45.5	19,738	31.2	14,766	23.3	63,316	100.0
Spruce	9,979	29.0	13,254	38.5	11,190	32.5	34,422	100.0
Birch	1,797	16.3	2,781	25.2	6,440	58.4	11,018	100.0
Other broadleaved	141	11.0	335	26.1	809	62.9	1,286	100.0
Total	40,729	37.0	36,109	32.8	33,204	30.2	110,042	100.0
North Finland								
Pine	36,682	52.5	20,736	29.7	12,459	17.8	69,878	100.0
Spruce	2,438	17.6	5,173	37.3	6,257	45.1	13,868	100.0
Birch	1,167	17.2	2,197	32.4	3,411	50.4	6,775	100.0
Other broadleaved	4	1.9	32	17.3	148	80.8	183	100.0
Total	40,291	44.4	28,138	31.0	22,276	24.6	90,705	100.0
Whole country								
Pine	65,495	49.2	40,475	30.4	27,225	20.4	133,194	100.0
Spruce	12,416	25.7	18,427	38.2	17,447	36.1	48,290	100.0
Birch	2,964	16.7	4,978	28.0	9,851	55.4	17,793	100.0
Other broadleaved	145	9.9	367	25.0	957	65.2	1,469	100.0
Total	81,020	40.4	64,247	32.0	55,480	27.6	200,747	100.0

Appendix

Table A.11 The areas of forest land based on the occurrance and proportion of most important second species[a] in the main storey on forest land

Second species	Proportion of the second species	Dominant tree species								Total	
		Pine		Spruce		Birch		Other broadleaved			
		km²	%	km²	%	km²	%	km²	%	km²	%
South Finland											
Scots pine	<25%	26	0.0	7,438	21.6	1,606	14.6	76	5.9	9,147	8.3
	≥25%	14	0.0	4,292	12.5	1,734	15.7	49	3.8	6,089	5.5
	Total	40	0.1	11,731	34.1	3,340	30.3	126	9.8	15,236	13.8
Norway spruce	<25%	10,587	16.7	.	.	1,524	13.8	136	10.6	12,247	11.1
	≥25%	5,364	8.5	.	.	1,499	13.6	98	7.6	6,962	6.3
	Total	15,952	25.2	.	.	3,023	27.4	234	18.2	19,209	17.5
Silver birch	<25%	6,072	9.6	3,702	10.8	563	5.1	169	13.1	10,505	9.5
	≥25%	1,550	2.4	1,100	3.2	395	3.6	117	9.1	3,162	2.9
	Total	7,622	12.0	4,801	13.9	958	8.7	286	22.2	13,667	12.4
Downy birch	<25%	7,347	11.6	4,911	14.3	523	4.8	161	12.5	12,942	11.8
	≥25%	3,204	5.1	1,973	5.7	637	5.8	195	15.2	6,009	5.5
	Total	10,551	16.7	6,883	20.0	1,160	10.5	356	27.7	18,951	17.2
Other coniferous	<25%	10	0.0	11	0.0	6	0.1	0	0.0	27	0.0
	≥25%	6	0.0	6	0.0	14	0.1	0	0.0	26	0.0
	Total	15	0.0	17	0.0	20	0.2	0	0.0	52	0.0
Other broadleaved	<25%	233	0.4	809	2.3	377	3.4	83	6.5	1,501	1.4
	≥25%	85	0.1	196	0.6	343	3.1	60	4.7	685	0.6
	Total	318	0.5	1,005	2.9	720	6.5	144	11.2	2,187	2.0
Total	<25%	24,274	38.3	16,871	49.0	4,598	41.7	625	48.6	46,369	42.1
	≥25%	10,224	16.1	7,567	22.0	4,622	42.0	520	40.4	22,932	20.8
	Total	34,498	54.5	24,438	71.0	9,221	83.7	1,145	89.0	69,301	63.0
Dominant tree species total		63,316	100.0	34,422	100.0	11,018	100.0	1,286	100.0	110,042	100.0

(continued)

Table A.11 (continued)

North Finland

Second species	Proportion of the second species	Pine km²	Pine %	Spruce km²	Spruce %	Birch km²	Birch %	Other broadleaved km²	Other broadleaved %	Total km²	Total %
Scots pine	<25%	11	0.0	3,258	23.5	1,657	24.5	14	7.7	4,939	5.4
	≥25%	21	0.0	2,120	15.3	1,309	19.3	21	11.6	3,471	3.8
	Total	32	0.0	5,378	38.8	2,965	43.8	35	19.3	8,410	9.3
Norway spruce	<25%	13,134	18.8	.	.	1,349	19.9	32	17.3	14,515	16.0
	≥25%	3,802	5.4	.	.	952	14.1	11	5.7	4,764	5.3
	Total	16,936	24.2	.	.	2,301	34.0	42	23.0	19,279	21.3
Silver birch	<25%	1,845	2.6	173	1.2	74	1.1	4	1.9	2,095	2.3
	≥25%	213	0.3	46	0.3	46	0.7	0	0.0	305	0.3
	Total	2,058	2.9	219	1.6	120	1.8	4	1.9	2,400	2.6
Downy birch	<25%	10,147	14.5	3,402	24.5	21	0.3	42	23.1	13,612	15.0
	≥25%	3,820	5.5	2,309	16.6	46	0.7	56	30.7	6,231	6.9
	Total	13,967	20.0	5,710	41.2	67	1.0	99	53.9	19,844	21.9
Other coniferous	<25%	32	0.0	0	0.0	4	0.1	0	0.0	35	0.0
	≥25%	7	0.0	14	0.1	0	0.0	0	0.0	21	0.0
	Total	39	0.1	14	0.1	4	0.1	0	0.0	56	0.1
Other broadleaved	<25%	77	0.1	74	0.5	67	1.0	0	0.0	219	0.2
	≥25%	34	0.0	25	0.2	84	1.2	0	0.0	143	0.2
	Total	111	0.2	99	0.7	151	2.2	0	0.0	362	0.4
Total	<25%	25,246	36.1	6,906	49.8	3,171	46.8	92	50.0	35,415	39.0
	≥25%	7,897	11.3	4,514	32.5	2,437	36.0	88	48.1	14,936	16.5
	Total	33,143	47.4	11,420	82.3	5,608	82.8	180	98.1	50,351	55.5
Dominant tree species total		69,878	100.0	13,868	100.0	6,775	100.0	183	100.0	90,705	100.0

Appendix

Whole country									
Scots pine	<25%	36	0.0	10,696	22.1	3,263	18.3	91	6.2	14,086	7.0
	≥25%	35	0.0	6,413	13.3	3,042	17.1	70	4.8	9,560	4.8
	Total	71	0.1	17,108	35.4	6,305	35.4	161	11.0	23,646	11.8
Norway spruce	<25%	23,722	17.8	.	.	2,873	16.1	167	11.4	26,762	13.3
	≥25%	9,166	6.9	.	.	2,452	13.8	109	7.4	11,726	5.8
	Total	32,888	24.7	.	.	5,324	29.9	276	18.8	38,488	19.2
Silver birch	<25%	7,916	5.9	3,874	8.0	637	3.6	173	11.7	12,600	6.3
	≥25%	1,764	1.3	1,146	2.4	441	2.5	117	7.9	3,467	1.7
	Total	9,680	7.3	5,020	10.4	1,077	6.1	289	19.7	16,067	8.0
Downey birch	<25%	17,494	13.1	8,312	17.2	545	3.1	203	13.8	26,554	13.2
	≥25%	7,024	5.3	4,282	8.9	683	3.8	252	17.1	12,240	6.1
	Total	24,518	18.4	12,594	26.1	1,227	6.9	455	31.0	38,794	19.3
Other coniferous	<25%	42	0.0	11	0.0	9	0.1	0	0.0	62	0.0
	≥25%	13	0.0	20	0.0	14	0.1	0	0.0	47	0.0
	Total	54	0.0	31	0.1	23	0.1	0	0.0	109	0.1
Other broadleaved	<25%	310	0.2	883	1.8	443	2.5	83	5.7	1,720	0.9
	≥25%	119	0.1	221	0.5	428	2.4	60	4.1	828	0.4
	Total	429	0.3	1,104	2.3	871	4.9	144	9.8	2,548	1.3
Total	<25%	49,520	37.2	23,777	49.2	7,769	43.7	717	48.8	81,784	40.7
	≥25%	18,121	13.6	12,081	25.0	7,059	39.7	608	41.4	37,868	18.9
	Total	67,641	50.8	35,858	74.3	14,829	83.3	1,324	90.1	119,652	59.6
Dominant tree species total		133,194	100.0	48,290	100.0	17,793	100.0	1,469	100.0	200,747	100.0

[a]Second species is the most important tree species after dominat tree species in the main storey considering to the wood production. No second species is recorded, if the proportion of dominant tree species is over 95%

Table A.12 The area of forest land and mean volume and mean diameter of the growing stock by dominant tree species groups and by age classes on forest land

(a) The area of forest land by age classes

Dominant tree species		Age class Temporarily unstocked	1–20	21–40	41–60	61–80	81–100	101–120	121–140	141–160	Over 160	Total
South Finland												
Pine	Area, km^2	.	11,701	16,266	9,869	8,451	7,866	5,242	2,563	944	415	63,316
	s.e., km^2	.	206	257	200	176	166	134	92	56	39	477
Spruce	Area, km^2	.	5,045	4,666	5,673	7,741	6,327	3,432	1,050	337	152	34,422
	s.e., km^2	.	131	125	133	158	146	108	58	33	21	355
Birch	Area, km^2	.	3,426	2,933	2,676	1,203	591	163	26	0	0	11,018
	s.e., km^2	.	104	96	88	59	42	22	9	.	.	197
Other broadleaved	Area, km^2	.	302	457	359	104	59	3	3	0	0	1,286
	s.e., km^2	.	30	37	34	17	14	3	3	.	.	63
Total	Area, km^2	1,629	20,474	24,322	18,578	17,499	14,842	8,839	3,642	1,280	566	111,671
	s.e., km^2	70	262	292	255	233	221	175	112	66	45	472
North Finland												
Pine	Area, km^2	.	9,981	10,590	11,248	13,085	8,904	3,993	2,786	1,929	7,363	69,878
	s.e., km^2	.	303	332	328	378	340	183	154	143	405	745
Spruce	Area, km^2	.	1,675	573	692	1,235	1,089	1,387	1,045	1,088	5,084	13,868
	s.e., km^2	.	113	64	65	86	80	106	88	89	312	427
Birch	Area, km^2	.	519	1,787	2,402	1,120	592	221	119	7	7	6,775
	s.e., km^2	.	53	116	138	89	71	50	32	7	7	235
Other broadleaved	Area, km^2	.	32	56	67	28	0	0	0	0	0	183
	s.e., km^2	.	12	16	19	11	.	.	.	.	.	29
Total	Area, km^2	1,002	12,207	13,007	14,409	15,468	10,585	5,601	3,949	3,024	12,454	91,707
	s.e., km^2	82	326	359	364	394	354	213	177	171	505	679

Appendix

Whole country											
Pine	Area, km²		21,682	26,856	21,117	21,536	16,770	9,235	5,349	2,872	7,777	133,194
	s.e., km²		366	419	384	417	378	226	179	154	407	885
Spruce	Area, km²		6,720	5,240	6,365	8,976	7,416	4,818	2,095	1,425	5,236	48,290
	s.e., km²		173	140	148	180	167	152	106	95	312	555
Birch	Area, km²		3,945	4,720	5,079	2,323	1,183	384	144	7	7	17,793
	s.e., km²		117	151	164	107	82	55	33	7	7	306
Other broadleaved	Area, km²		333	513	426	133	59	3	3	0	0	1,469
	s.e., km²		32	40	39	20	14	3	3	.	.	70
Total	Area, km²	2,631	32,681	37,328	32,987	32,968	25,427	14,440	7,591	4,304	13,020	203,378
	s.e., km²	108	418	463	444	458	418	276	210	184	507	827

(b) The mean volume of growing stock on forest land by age classes

South Finland

Pine	m³/ha		21.5	84.3	112.8	144.3	166.1	168.3	159.5	156.0	161.5	106.9
	s.e., m³/ha		0.5	0.8	1.3	1.6	1.9	2.4	3.2	5.5	8.9	0.7
Spruce	m³/ha		22.9	103.8	173.9	221.3	237.1	238.7	232.3	220.5	200.2	173.4
	s.e., m³/ha		0.8	1.8	2.0	1.9	2.2	2.9	5.4	8.9	17.8	1.1
Birch	m³/ha		18.9	89.7	134.7	164.9	192.8	202.1	165.2	–	–	94.2
	s.e., m³/ha		0.8	1.8	2.6	4.2	7.5	13.2	66.8	–	–	1.4
Other broadleaved	m³/ha		55.3	132.4	187.8	247.7	206.9	198.0	255.4	–	–	142.9
	s.e., m³/ha		5.9	6.9	10.1	22.5	33.3	198.0	255.4	–	–	5.1
Total	m³/ha	7.6	21.9	89.6	136.1	180.4	197.6	196.3	180.7	173.0	171.9	125.1
	s.e., m³/ha	1.0	0.4	0.7	1.0	1.3	1.6	2.0	2.9	4.9	8.1	0.6

North Finland

Pine	m³/ha		13.0	50.0	61.7	71.2	82.9	97.4	109.4	114.4	92.6	66.1
	s.e., m³/ha		0.5	0.6	0.8	0.5	1.5	1.4	1.9	4.7	1.5	0.6
Spruce	m³/ha		10.5	38.3	70.5	98.1	117.0	115.2	124.4	125.0	101.0	92.0
	s.e., m³/ha		0.9	3.3	4.1	3.4	2.7	1.3	0.4	1.2	1.2	1.1

(continued)

Table A.12 (continued)

Dominant tree species		Age class (year) Temporarily unstocked	1–20	21–40	41–60	61–80	81–100	101–120	121–140	141–160	Over 160	Total
Birch	m³/ha	.	15.8	60.9	77.2	87.8	83.5	83.0	72.3	37.4	105.1	70.6
	s.e., m³/ha		1.8	1.8	0.8	1.5	2.9	5.1	5.0	37.4	105.1	1.0
Other broadleaved	m³/ha	.	28.2	94.4	121.9	134.8	–	–	–	–	–	99.3
	s.e., m³/ha		12.4	9.8	19.3	26.4						9.3
Total	m³/ha	8.6	12.8	51.2	65.0	74.6	86.5	101.2	112.3	118.1	96.0	69.8
	s.e., m³/ha	1.4	0.4	0.6	0.6	0.3	1.4	1.2	1.5	3.5	1.1	0.6
Whole country												
Pine	m³/ha	.	17.6	70.8	85.6	99.9	122.0	137.7	133.4	128.1	96.2	85.5
	s.e., m³/ha		0.3	0.3	0.4	0.8	1.2	0.7	0.8	3.6	0.8	0.4
Spruce	m³/ha	.	19.8	96.7	162.6	204.4	219.4	203.1	178.5	147.6	103.8	150.0
	s.e., m³/ha		0.6	1.5	1.7	1.5	1.7	3.0	4.6	5.3	3.3	0.2
Birch	m³/ha	.	18.5	78.8	107.5	127.7	138.1	133.4	88.8	37.4	105.1	85.2
	s.e., m³/ha		0.8	1.0	1.6	3.0	5.3	12.0	9.7	37.4	105.1	0.9
Other broadleaved	m³/ha	.	52.7	128.2	177.4	223.7	206.9	198.0	255.4	–	–	137.5
	s.e., m³/ha	.	5.5	6.1	8.5	17.3	33.3	198.0	255.4			4.6
Total	m³/ha	8.0	18.5	76.2	105.0	130.8	151.3	159.4	145.1	134.4	99.3	100.2
	s.e., m³/ha	0.8	0.3	0.3	0.1	0.7	1.0	0.6	0.7	2.9	0.7	0.4

204 Appendix

(c) Mean diameter of growing stock on forest land by age classes (cm)

South Finland										
Pine	.	10.3	13.3	16.5	20.3	24.0	25.8	26.2	26.2	28.7	17.4
Spruce	.	13.4	14.4	19.7	25.3	27.8	28.1	27.1	26.9	25.6	22.0
Birch	.	10.7	12.9	17.0	21.4	25.7	29.7	37.7	.	.	15.1
Other broadleaved	.	15.2	14.1	20.5	29.9	25.8	33.8	18.1	.	.	18.0
Total	17.4	11.2	13.5	17.7	22.7	25.7	26.8	26.5	26.4	27.9	18.6
North Finland											
Pine	.	12.1	13.1	14.4	15.4	17.7	20.8	23.1	26.3	33.3	17.5
Spruce	.	10.8	10.2	14.5	15.8	17.8	18.8	21.8	22.9	28.2	20.9
Birch	.	7.2	10.5	14.1	15.0	17.0	16.9	17.4	12.6	17.0	13.2
Other broadleaved	.	8.6	10.4	14.6	23.2	.	.	.	.	.	13.6
Total	18.3	11.7	12.6	14.4	15.4	17.6	20.2	22.6	25.0	31.2	17.7
Whole country											
Pine	.	11.2	13.2	15.4	17.3	20.6	23.7	24.6	26.3	33.0	17.5
Spruce	.	12.7	14.0	19.2	24.0	26.3	25.4	24.5	23.8	28.1	21.7
Birch	.	10.3	12.0	15.6	18.3	21.3	22.3	21.0	12.6	17.0	14.4
Other broadleaved	.	14.6	13.7	19.6	28.4	25.8	33.8	18.1	.	.	17.5
Total	17.7	11.4	13.2	16.2	19.3	22.3	24.2	24.5	25.4	31.0	18.2

Table A.13 The areas of forest land and mean characteristics of growing stock by development classes on forest land (a) and on forest land available for wood production (b)

	Development class								Total
	1	2	3	4	5	6	7	8	
(a) Forest land total									
South Finland									
Area, km²	1,629	7,404	16,643	37,505	30,076	16,692	427	1,295	111,671
Capable of development	1,380	7,123	15,757	35,696	29,436	15,244	277	1,111	106,023
Low-yielding	249	281	886	1,809	640	1,448	151	184	5,648
Mean age, year	0	5	19	43	74	110	104	106	56
Basal area, m²/ha	1	2	6	17	24	27	13	4	17
Mean diameter, cm									
Pine	17	30	11	15	23	30	32	31	21
Spruce	10	15	9	14	23	26	27	13	21
Birch	18	14	8	12	18	22	24	20	15
Other broadleaved	16	11	8	11	15	17	15	15	13
Total	16	20	9	14	22	27	28	26	20
Mean volume, m³/ha									
Pine	1	6	11	50	68	93	55	21	51
Spruce	1	2	4	25	91	108	31	3	50
Birch	3	3	8	19	24	26	17	5	18
Other broadleaved	3	1	4	5	7	7	5	1	5
Total	8	11	27	99	190	234	107	30	125
Timing of proposed cutting, % of the area									
Delayed	1	3	11	13	11	11	16	2	11
First 5-year period	0	35	35	24	25	68	36	24	33
Second 5-year period	0	33	18	27	20	15	20	31	22
No cuttings	99	29	36	36	44	6	27	44	34

Appendix

North Finland								
Area, km²	1,002	5,422	13,325	38,047	15,809	16,640	1,223	91,707
Capable of development	790	5,093	12,332	34,324	14,273	7,967	239	75,650
Low-yielding	212	329	992	3,723	1,535	8,673	70	16,057
Mean age, year	0	6	24	58	91	180	169	76
Basal area, m²/ha	2	2	5	13	17	17	179	12
Mean diameter, cm							11	
Pine	19	23	10	15	20	27	27	18
Spruce	11	12	9	13	18	23	14	19
Birch	11	10	6	10	13	15	10	11
Other broadleaved	15	22	9	10	14	19	.	14
Total	15	18	9	13	18	23	25	17
Mean volume, m³/ha								
Pine	4	7	13	43	66	59	60	43
Spruce	2	1	2	7	22	40	8	14
Birch	2	2	5	13	17	14	4	12
Other broadleaved	1	0	1	1	2	2	0	1
Total	9	10	20	64	107	114	72	70
Timing of proposed cutting, % of the area								
Delayed	1	2	9	15	13	48	61	19
First 5-year period	0	19	23	13	17	36	7	19
Second 5-year period	0	24	15	16	15	4	15	14
No cuttings	99	55	53	56	55	12	16	48

(continued)

Table A.13 (continued)

	Development class								Total
	1	2	3	4	5	6	7	8	
Whole country									
Area, km²	2,631	12,826	29,968	75,552	45,884	33,332	667	2,518	203,378
Capable of development	2,170	12,216	28,089	70,020	43,709	23,211	346	1,910	181,673
Low-yielding	461	610	1,879	5,532	2,175	10,121	320	608	21,705
Mean age, year	0	5	20	48	78	132	117	122	62
Basal area, m²/ha	1	2	6	15	22	22	12	4	15
Mean diameter, cm									
Pine	18	27	11	15	22	29	30	29	20
Spruce	11	14	9	13	21	24	23	11	20
Birch	15	13	7	11	16	18	19	15	13
Other broadleaved	16	16	8	11	15	18	10	12	14
Total	15	19	9	13	21	25	27	25	18
Mean volume, m³/ha									
Pine	2	6	12	46	67	76	56	20	48
Spruce	2	1	3	16	67	74	22	2	34
Birch	2	2	7	16	22	20	13	3	15
Other broadleaved	2	1	2	3	5	4	3	1	3
Total	8	11	24	81	162	174	95	26	100
Timing of proposed cutting, % of the area									
Delayed	1	3	10	14	12	29	32	5	14
First 5-year period	0	28	29	18	22	52	26	15	27
Second 5-year period	0	29	17	22	18	10	18	28	18
No cuttings	99	40	44	46	48	9	23	52	40

Appendix

(b) Forest land available for wood production

South Finland

Area, km²	1,620	7,384	16,499	37,128	29,794	16,232	427	1,295	110,379
Capable of development	1,377	7,103	15,624	35,339	29,174	14,942	277	1,111	104,946
Low-yielding	243	281	875	1789	620	1,290	151	184	5,433
Mean age, year	.	5	19	43	74	109	104	106	56
Basal area, m²/ha	1	2	6	17	24	27	13	4	17
Mean diameter, cm									
Pine	17	30	11	15	23	30	32	31	21
Spruce	10	15	9	14	23	26	27	13	21
Birch	18	14	8	12	18	22	24	20	15
Other broadleaved	16	11	8	11	15	17	15	15	13
Total	16	20	9	14	22	27	28	26	19
Mean volume, m³/ha									
Pine	1	6	11	50	68	93	55	21	51
Spruce	1	2	4	25	92	109	31	3	50
Birch	3	3	8	19	24	26	17	5	18
Other broadleaved	3	1	4	5	7	7	5	1	5
Total	8	11	27	99	190	234	107	30	125
Timing of proposed cutting, % of the area									
Delayed	1	3	11	13	11	11	16	2	11
First 5-year period	0	35	35	24	25	69	36	24	33
Second 5-year period	0	33	18	27	20	16	20	31	22
No cuttings	99	28	36	36	44	5	27	44	34

(continued)

Table A.13 (continued)

	Development class								Total
	1	2	3	4	5	6	7	8	
North Finland									
Area, km^2	992	5,330	13,032	36,511	14,319	10,057	232	1,188	81,662
Capable of development	787	5,015	12,101	33,040	13,182	5,838	66	796	70,826
Low-yielding	205	315	931	3,471	1,138	4,219	166	391	10,836
Mean age, year	.	6	24	58	90	167	180	166	68
Basal area, m^2/ha	2	2	5	13	18	18	11	3	12
Mean diameter, cm									
Pine	19	23	10	14	20	27	26	28	18
Spruce	11	12	9	13	17	21	14	10	17
Birch	11	10	6	10	13	15	9	10	11
Other broadleaved	15	21	9	11	14	18	.	9	14
Total	15	18	9	13	18	23	24	24	16
Mean volume, m^3/ha									
Pine	4	7	12	43	69	64	58	19	42
Spruce	2	1	2	7	22	40	8	1	12
Birch	2	2	5	13	18	15	5	2	12
Other broadleaved	1	0	1	1	2	2	0	1	1
Total	9	10	20	64	111	122	71	22	67
Timing of proposed cutting, % of the area									
Delayed	1	3	9	15	13	46	61	8	16
First 5-year period	0	19	23	13	18	46	8	5	20
Second 5-year period	0	24	15	17	16	7	15	25	15
No cuttings	99	55	53	56	53	1	15	62	48

Appendix

Whole country								
Area, km²	2,612	12,714	29,531	73,640	44,113	26,289	660	2,482	192,041
Capable of development	2,164	12,119	27,725	68,379	42,355	20,780	343	1,907	175,772
Low-yielding	448	595	1,806	5,261	1,758	5,509	317	575	16,269
Mean age, year	0	5	20	48	77	123	116	122	60
Basal area, m²/ha	1	2	6	15	22	23	12	4	15
Mean diameter, cm									
Pine	18	27	10	14	22	29	30	29	19
Spruce	10	13	9	13	21	24	23	11	20
Birch	15	12	7	11	16	19	19	15	13
Other broadleaved	16	16	9	11	15	17	10	12	13
Total	15	19	9	13	21	25	27	25	18
Mean volume, m³/ha									
Pine	2	6	12	46	68	82	56	20	47
Spruce	2	1	3	16	69	82	23	2	34
Birch	2	2	7	16	22	22	13	4	16
Other broadleaved	2	1	2	3	5	5	3	1	4
Total	8	11	24	81	164	191	95	26	100
Timing of proposed cutting, % of the area									
Delayed	1	3	10	14	12	24	32	5	13
First 5-year period	0	28	30	19	23	60	26	15	27
Second 5-year period	0	30	17	22	19	12	19	28	19
No cuttings	99	39	44	46	47	3	23	53	40

Development classes: *1* Temporarily unstocked regeneration area, *2* Young seedling stand, *3* Advanced seedling stand, *4* Young thinning stand, *5* Advanced thinning stand, *6* Mature stand, *7* Shelter tree stand, *8* Seed tree stand

Table A.14 The area of forest land by tree species dominance and by tree storeys

Dominant tree species of the tree storey	Under storey Development capable	Non established seedlings	Not development capable	Total	Proportion of forest land area	Upper storey Over storey	Retention tree storey	Nurse crop	Total	Proportion of forest land area
	km²				%	km²				%
South Finland										
Pine	465	115	264	844	0.8	1,743	602	0	2,345	2.1
Spruce	1,456	521	2,622	4,600	4.1	182	62	0	244	0.2
Broadleaved	76	6	2,708	2,789	2.5	861	303	570	1,735	1.6
Forest land total	1,998	641	5,594	8,233	7.4	2,786	968	570	4,324	3.9
North Finland										
Pine	800	373	401	1,575	1.7	2,630	1,433	0	4,063	4.4
Spruce	496	64	448	1,008	1.1	71	64	0	136	0.1
Broadleaved	7	0	1,902	1,909	2.1	212	46	64	322	0.4
Forest land total	1,303	437	2,752	4,492	4.9	2,914	1,543	64	4,521	4.9
Whole country										
Pine	1,265	488	666	2,419	1.2	4,373	2,035	0	6,408	3.2
Spruce	1,952	585	3,071	5,607	2.8	253	126	0	380	0.2
Broadleaved	83	6	4,610	4,699	2.3	1,074	349	634	2,057	1.0
Forest land total	3,301	1,078	8,347	12,725	6.3	5,700	2,511	634	8,845	4.3

Table A.15 The volume of the growing stock

(a) The volume of the growing stock by tree species on forest land and on poorly productive forest land

	Forest land				Poorly productive forest land				Total				
	Mean volume m³/ha	s.e. m³/ha	Total volume 1,000 m³	s.e. 1,000 m³	%	Mean volume m³/ha	s.e. m³/ha	Total volume 1,000 m³	s.e. 1,000 m³	%	Mean volume m³/ha	s.e. m³/ha	Total volume 1,000 m³	s.e. 1,000 m³	%
South Finland															
Pine	51.4	0.4	574,421	5,017	41.1	20.8	0.6	8,908	370	83.7	50.3	0.4	583,329	5,018	41.4
Norway spruce	49.9	0.5	557,189	5,932	39.9	1.3	0.2	553	76	5.2	48.1	0.5	557,742	5,968	39.6
Silver birch	6.0	0.1	67,226	1,232	4.8	0.4	0.1	189	32	1.8	5.8	0.1	67,416	1,235	4.8
Downy birch	12.4	0.1	138,464	1,762	9.9	1.4	0.1	610	65	5.7	12.0	0.1	139,074	1,770	9.9
Other broadleaved	5.3	0.1	59,616	1,315	4.3	0.9	0.2	382	81	3.6	5.2	0.1	59,998	1,320	4.3
Total	125.1	0.6	1,396,917	8,841	100.0	24.9	0.7	10,642	435	100.0	121.4	0.6	1,407,559	8,887	100.0
North Finland															
Pine	42.7	0.5	391,651	5,561	61.2	11.0	0.4	24,661	1,079	56.6	36.5	0.4	416,312	5,575	60.9
Norway spruce	14.2	0.3	129,960	3,333	20.3	3.1	0.2	6,980	536	16.0	12.0	0.3	136,940	3,446	20.0
Silver birch	0.8	0.0	7,739	389	1.2	0.1	0.0	113	30	0.3	0.7	0.0	7,852	389	1.1
Downy birch	10.8	0.2	99,344	1,892	15.5	5.1	0.4	11,496	872	26.4	9.7	0.2	110,840	2,080	16.2
Other broadleaved	1.2	0.1	11,344	533	1.8	0.2	0.0	347	78	0.8	1.0	0.0	11,690	538	1.7
Total	69.8	0.6	640,037	6,979	100.0	19.4	0.5	43,597	1,507	100.0	59.9	0.5	683,634	6,759	100.0
Whole country															
Pine	47.5	0.3	966,072	7,489	47.4	12.6	0.3	33,569	1,141	61.9	43.4	0.3	999,640	7,501	47.8
Norway spruce	33.8	0.3	687,149	6,804	33.7	2.8	0.2	7,533	542	13.9	30.2	0.3	694,682	6,891	33.2
Silver birch	3.7	0.1	74,965	1,292	3.7	0.1	0.0	303	43	0.6	3.3	0.1	75,268	1,294	3.6
Downy birch	11.7	0.1	237,808	2,586	11.7	4.5	0.3	12,106	874	22.3	10.9	0.1	249,914	2,731	12.0
Other broadleaved	3.5	0.1	70,960	1,419	3.5	0.3	0.0	729	112	1.3	3.1	0.1	71,689	1,425	3.4
Total	100.2	0.4	2,036,954	11,264	100.0	20.3	0.4	54,239	1,568	100.0	90.9	0.4	2,091,193	11,165	100.0

(continued)

Table A.15 (continued)
(b) The volume of the growing stock by principal site classes on forest land and on poorly productive forest land

	Forest land			Poorly productive forest land			Total		
	Mean volume, m³/ha	Total volume, 1,000 m³	%	Mean volume, m³/ha	Total volume, 1,000 m³	%	Mean volume, m³/ha	Total volume, 1,000 m³	%
South Finland									
Mineral soil									
Undrained	133.5	1,021,722		38.2	6,104		131.5	1,027,827	
Drained	110.7	93,299		–	–		110.7	93,299	
Pine	52.4	445,320	39.9	32.0	5,109	83.7	52.0	450,430	40.2
Norway spruce	55.8	473,932	42.5	2.6	420	6.9	54.8	474,352	42.3
Silver birch	7.3	62,396	5.6	1.2	189	3.1	7.2	62,586	5.6
Downy birch	9.3	78,973	7.1	0.9	149	2.4	9.1	79,122	7.1
Other broadleaved	6.4	54,400	4.9	1.5	237	3.9	6.3	54,636	4.9
Total	131.2	1,115,021	100.0	38.2	6,104	100.0	129.5	1,121,125	100.0
Peatland									
Undrained	108.5	35,419		20.3	2,663		83.3	38,081	
Drained	105.2	246,477		13.7	1,875		100.1	248,352	
Pine	48.4	129,101	45.8	14.2	3,798	83.7	45.2	132,899	46.4
Norway spruce	31.2	83,257	29.5	0.5	133	2.9	28.4	83,390	29.1
Silver birch	1.8	4,830	1.7	0.0	0	0.0	1.6	4,830	1.7
Downy birch	22.3	59,492	21.1	1.7	461	10.2	20.4	59,953	20.9
Other broadleaved	2.0	5,217	1.9	0.5	145	3.2	1.8	5,362	1.9
Total	105.6	281,896	100.0	16.9	4,538	100.0	97.5	286,433	100.0

Appendix

North Finland							
Mineral soil							
Undrained	71.0	440,652		23.3	17,109	65.9	457,766
Drained	68.5	33,268		0.0	0	68.4	33,268
Pine	45.6	305,125	64.4	9.5	6,963	42.0	312,094	63.6
Norway spruce	14.8	98,878	20.9	3.2	2,322	13.6	101,200	20.6
Silver birch	1.1	7,283	1.5	0.1	97	1.0	7,379	1.5
Downy birch	8.0	53,482	11.3	10.3	7,572	8.2	61,054	12.4
Other broadleaved	1.4	9,152	1.9	0.2	155	1.3	9,307	1.9
Total	70.8	473,920	100.0	23.3	17,109	66.1	491,034	100.0
Peatland								
Undrained	63.4	32,938		18.7	20,978	32.8	53,916	
Drained	68.0	133,179		14.3	5,505	59.2	138,684	
Pine	34.9	86,525	52.1	11.7	17,692	26.1	104,217	54.1
Norway spruce	12.5	31,082	18.7	3.1	4,658	9.0	35,740	18.6
Silver birch	0.2	456	0.3	0.0	16	0.1	473	0.2
Downy birch	18.5	45,862	27.6	2.6	3,924	12.5	49,786	25.8
Other broadleaved	0.9	2,192	1.3	0.1	192	0.6	2,384	1.2
Total	67.0	166,118	100.0	17.6	26,482	48.3	192,600	100.0
Whole country								
Mineral soil								
Undrained	105.5	1,462,374		25.9	23,213	100.7	1,485,593	
Drained	95.3	126,567		0.0	0	95.3	126,567	
Pine	49.4	750,446	47.2	13.5	12,072	47.4	762,524	47.3
Norway spruce	37.7	572,810	36.0	3.1	2,741	35.8	575,552	35.7
Silver birch	4.6	69,679	4.4	0.3	286	4.3	69,965	4.3
Downy birch	8.7	132,455	8.3	8.6	7,721	8.7	140,176	8.7
Other broadleaved	4.2	63,551	4.0	0.4	392	4.0	63,943	4.0
Total	104.6	1,588,941	100.0	25.9	23,213	100.2	1,612,160	100.0

(continued)

Table A.15 (continued)

Peatland	Forest land Mean volume, m³/ha	Forest land Total volume, 1,000 m³	%	Poorly productive forest land Mean volume, m³/ha	Poorly productive forest land Total volume, 1,000 m³	%	Total Mean volume, m³/ha	Total Total volume, 1,000 m³	%
Undrained	80.8	68,357		18.9	23,640		43.8	91,997	
Drained	88.2	379,656		14.1	7,380		80.2	387,036	
Pine	41.9	215,626	48.1	12.1	21,490	69.3	34.2	237,116	49.5
Norway spruce	22.2	114,339	25.5	2.7	4,792	15.4	17.2	119,131	24.9
Silver birch	1.0	5,286	1.2	0.0	16	0.1	0.8	5,303	1.1
Downy birch	20.5	105,353	23.5	2.5	4,385	14.1	15.9	109,738	22.9
Other broadleaved	1.4	7,409	1.7	0.2	337	1.1	1.1	7,746	1.6
Total	87.0	448,013	100.0	17.5	31,020	100.0	69.2	479,033	100.0

(c) The volume of the growing stock by tree species on forest land and on poorly productive forest land available for wood production

	Forest land					Poorly productive forest land					Total				
	Mean volume m³/ha	s.e. m³/ha	Total volume 1,000 m³	s.e. 1,000 m³	Proportion of tree species %	Mean volume m³/ha	s.e. m³/ha	Total volume 1,000 m³	s.e. 1,000 m³	Proportion of tree species %	Mean volume m³/ha	s.e. m³/ha	Total volume 1,000 m³	s.e. 1,000 m³	Proportion of tree species %
South Finland															
Pine	51.1	0.4	564,324	5,012	41.0	21.1	0.7	8,221	356	83.4	50.1	0.4	572,545	5,027	41.3
Norway spruce	49.9	0.5	551,313	5,934	40.1	1.4	0.2	538	76	5.5	48.3	0.5	551,851	5,973	39.8
Silver birch	6.0	0.1	65,997	1,224	4.8	0.5	0.1	187	32	1.9	5.8	0.1	66,185	1,226	4.8
Downy birch	12.4	0.1	136,680	1,759	9.9	1.4	0.2	549	61	5.6	12.0	0.1	137,229	1,769	9.9
Other broadleaved	5.3	0.1	58,097	1,289	4.2	0.9	0.2	361	80	3.7	5.1	0.1	58,458	1,293	4.2
Total	124.7	0.6	1,376,411	8,902	100.0	25.3	0.8	9,857	420	100.0	121.3	0.6	1,386,267	8,975	100.0
North Finland															
Pine	41.9	0.5	342,397	5,134	62.3	11.7	0.4	17,847	834	65.0	37.2	0.4	360,244	5,217	62.4
Norway spruce	12.4	0.3	101,017	2,545	18.4	2.6	0.2	3,892	291	14.2	10.8	0.2	104,909	2,597	18.2
Silver birch	0.8	0.0	6,483	341	1.2	0.1	0.0	93	28	0.3	0.7	0.0	6,577	342	1.1
Downy birch	10.9	0.2	89,362	1,797	16.3	3.5	0.4	5,378	598	19.6	9.8	0.2	94,740	1,897	16.4
Other broadleaved	1.3	0.1	10,365	495	1.9	0.2	0.0	243	70	0.9	1.1	0.1	10,608	499	1.8
Total	67.3	0.5	549,625	6,563	100.0	18.1	0.5	27,453	1,160	100.0	59.6	0.4	577,078	6,644	100.0
Whole country															
Pine	47.2	0.3	906,721	7,175	47.1	13.7	0.3	26,068	907	69.9	44.2	0.3	932,789	7,245	47.5
Norway spruce	34.0	0.3	652,330	6,457	33.9	2.3	0.1	4,430	300	11.9	31.1	0.3	656,760	6,513	33.5
Silver birch	3.8	0.1	72,481	1,271	3.8	0.1	0.0	281	43	0.8	3.4	0.1	72,761	1,273	3.7
Downy birch	11.8	0.1	226,042	2,514	11.7	3.1	0.3	5,927	601	15.9	11.0	0.1	231,969	2,594	11.8
Other broadleaved	3.6	0.1	68,461	1,381	3.6	0.3	0.1	604	106	1.6	3.3	0.1	69,065	1,386	3.5
Total	100.3	0.3	1,926,036	11,060	100.0	19.5	0.4	37,310	1,234	100.0	93.0	0.3	1,963,345	11,167	100.0

(continued)

Table A.15 (continued)

(d) The volume of the growing stock by tree species on FAO forest and on FAO other wooded land

	Forest					Other wooded land					Total				
	Mean volume m³/ha	s.e. m³/ha	Total volume 1,000 m³	s.e. 1,000 m³	Proportion of tree species %	Mean volume m³/ha	s.e. m³/ha	Total volume 1,000 m³	s.e. 1,000 m³	Proportion of tree species %	Mean volume m³/ha	s.e. m³/ha	Total volume 1,000 m³	s.e. 1,000 m³	Proportion of tree species %
South Finland															
Pine	50.5	0.4	581,954	5,036	41.4	10.8	0.7	1,180	103	87.6	50.1	0.4	583,134	5,031	41.4
Norway spruce	48.4	0.5	557,689	5,962	39.7	0.4	0.2	47	24	3.5	47.9	0.5	557,736	5,972	39.6
Silver birch	5.8	0.1	67,404	1,234	4.8	0.1	0.1	11	6	0.9	5.8	0.1	67,416	1,235	4.8
Downy birch	12.1	0.1	138,981	1,766	9.9	0.8	0.2	87	21	6.5	11.9	0.1	139,068	1,768	9.9
Other broadleaved	5.2	0.1	59,963	1,320	4.3	0.2	0.1	20	8	1.5	5.2	0.1	59,984	1,320	4.3
Total	121.9	0.6	1,405,991	8,904	100.0	12.3	0.9	1,347	118	100.0	120.9	0.6	1,407,338	8,908	100.0
North Finland															
Pine	37.7	0.5	413,268	5,572	60.8	3.8	0.2	2,712	215	69.1	35.6	0.4	415,980	5,576	60.9
Norway spruce	12.5	0.3	136,496	3,439	20.1	0.6	0.1	405	81	10.3	11.7	0.3	136,901	3,440	20.0
Silver birch	0.7	0.0	7,850	389	1.2	0.0	0.0	2	2	0.0	0.7	0.0	7,852	389	1.1
Downy birch	10.0	0.2	109,876	2,079	16.2	1.1	0.2	795	129	20.3	9.5	0.2	110,671	2,039	16.2
Other broadleaved	1.1	0.0	11,679	538	1.7	0.0	0.0	12	9	0.3	1.0	0.0	11,690	538	1.7
Total	62.0	0.5	679,169	6,754	100.0	5.5	0.3	3,926	286	100.0	58.5	0.5	683,095	6,713	100.0
Whole country															
Pine	44.3	0.3	995,223	7,511	47.7	4.7	0.2	3,892	238	73.8	42.9	0.3	999,115	7,510	47.8
Norway spruce	30.9	0.3	694,185	6,883	33.3	0.5	0.1	453	84	8.6	29.8	0.3	694,637	6,891	33.2
Silver birch	3.3	0.1	75,255	1,294	3.6	0.0	0.0	13	6	0.3	3.2	0.1	75,268	1,295	3.6
Downy birch	11.1	0.1	248,857	2,728	11.9	1.1	0.2	882	131	16.7	10.7	0.1	249,739	2,699	11.9
Other broadleaved	3.2	0.1	71,642	1,425	3.4	0.0	0.0	32	12	0.6	3.1	0.1	71,674	1,425	3.4
Total	92.7	0.4	2,085,160	11,175	100.0	6.4	0.3	5,273	310	100.0	89.7	0.4	2,090,433	11,154	100.0

Table A.16 The volume of the growing stock by timber assortment

		Saw-timber		Pulpwood		Waste wood		Total	
	Tree species	1,000 m^3	%	1,000 m^3	%	1,000 m^3	%	1,000 m^3	%

(a) The volume of the growing stock by principal site classes and by tree species on forest land and poorly productive forest land

South Finland

Mineral soil	Pine	179,556	39.9	250,395	55.6	20,479	4.5	450,430	100.0
	Norway spruce	225,094	47.5	225,919	47.6	23,338	4.9	474,352	100.0
	Silver birch	16,418	26.2	41,003	65.5	5,166	8.3	62,586	100.0
	Downy birch	8,089	10.2	56,226	71.1	14,807	18.7	79,122	100.0
	Other broadleaved	4,881	8.9	34,759	63.6	14,997	27.4	54,636	100.0
	Total	434,037	38.7	608,301	54.3	78,787	7.0	1,121,125	100.0
Peatland	Pine	32,010	24.1	91,124	68.6	9,765	7.3	132,899	100.0
	Norway spruce	32,486	39.0	44,947	53.9	5,958	7.1	83,390	100.0
	Silver birch	1,044	21.6	3,354	69.4	433	9.0	4,830	100.0
	Downy birch	3,628	6.1	44,328	73.9	11,996	20.0	59,953	100.0
	Other broadleaved	482	9.0	3,489	65.1	1,391	25.9	5,362	100.0
	Total	69,650	24.3	187,241	65.4	29,543	10.3	286,433	100.0
Total	Pine	211,566	36.3	341,519	58.5	30,244	5.2	583,329	100.0
	Norway spruce	257,580	46.2	270,866	48.6	29,296	5.3	557,742	100.0
	Silver birch	17,461	25.9	44,356	65.8	5,598	8.3	67,416	100.0
	Downy birch	11,717	8.4	100,554	72.3	26,804	19.3	139,074	100.0
	Other broadleaved	5,363	8.9	38,248	63.7	16,388	27.3	59,998	100.0
	Total	503,687	35.8	795,542	56.5	108,330	7.7	1,407,559	100.0

North Finland

Mineral soil	Pine	76,383	24.5	216,959	69.5	18,746	6.0	312,088	100.0
	Norway spruce	30,069	29.7	63,731	63.0	7,400	7.3	101,200	100.0
	Silver birch	479	6.5	6,014	81.5	886	12.0	7,379	100.0
	Downy birch	494	0.8	39,117	64.1	21,443	35.1	61,054	100.0
	Other broadleaved	391	4.2	5,676	61.0	3,239	34.8	9,307	100.0
	Total	107,816	22.0	331,498	67.5	51,714	10.5	491,028	100.0

(continued)

Table A.16 (continued)

	Tree species	Saw-timber 1,000 m³	%	Pulpwood 1,000 m³	%	Waste wood 1,000 m³	%	Total 1,000 m³	%
Peatland	Pine	9,587	9.2	80,786	77.5	13,850	13.3	104,224	100.0
	Norway spruce	5,400	15.1	25,555	71.5	4,785	13.4	35,740	100.0
	Silver birch	21	4.5	359	76.0	92	19.5	473	100.0
	Downy birch	379	0.8	35,278	70.9	14,128	28.4	49,786	100.0
	Other broadleaved	45	1.9	1,286	54.0	1,052	44.1	2,384	100.0
	Total	15,433	8.0	143,265	74.4	33,908	17.6	192,606	100.0
Total	Pine	85,970	20.7	297,745	71.5	32,597	7.8	416,312	100.0
	Norway spruce	35,469	25.9	89,286	65.2	12,185	8.9	136,940	100.0
	Silver birch	501	6.4	6,373	81.2	978	12.5	7,852	100.0
	Downy birch	873	0.8	74,395	67.1	35,571	32.1	110,840	100.0
	Other broadleaved	436	3.7	6,963	59.6	4,291	36.7	11,690	100.0
	Total	123,249	18.0	474,762	69.4	85,623	12.5	683,634	100.0
Whole country									
Mineral soil	Pine	255,939	33.6	467,354	61.3	39,225	5.1	762,518	100.0
	Norway spruce	255,163	44.3	289,651	50.3	30,738	5.3	575,552	100.0
	Silver birch	16,897	24.2	47,017	67.2	6,052	8.6	69,965	100.0
	Downy birch	8,583	6.1	95,343	68.0	36,250	25.9	140,176	100.0
	Other broadleaved	5,272	8.2	40,435	63.2	18,236	28.5	63,943	100.0
	Total	541,854	33.6	939,799	58.3	130,501	8.1	1,612,154	100.0
Peatland	Pine	41,597	17.5	171,910	72.5	23,616	10.0	237,122	100.0
	Norway spruce	37,886	31.8	70,502	59.2	10,743	9.0	119,131	100.0
	Silver birch	1,065	20.1	3,713	70.0	525	9.9	5,303	100.0
	Downy birch	4,008	3.7	79,606	72.5	26,125	23.8	109,738	100.0
	Other broadleaved	527	6.8	4,776	61.7	2,443	31.5	7,746	100.0
	Total	85,083	17.8	330,506	69.0	63,451	13.2	479,039	100.0

Total	Pine	297,536	29.8	639,264	63.9	62,841	6.3	999,640	100.0
	Norway spruce	293,049	42.2	360,152	51.8	41,481	6.0	694,682	100.0
	Silver birch	17,962	23.9	50,730	67.4	6,577	8.7	75,268	100.0
	Downy birch	12,590	5.0	174,949	70.0	62,375	25.0	249,914	100.0
	Other broadleaved	5,799	8.1	45,211	63.1	20,679	28.8	71,689	100.0
	Total	626,936	30.0	1,270,305	60.7	193,952	9.3	2,091,193	100.0

(b) The volume of the growing stock by tree species and by timber assortment on forest land and poorly productive forest land available for wood production

South Finland	Pine	207,230	36.2	335,630	58.6	29,685	5.2	572,545	100.0
	Norway spruce	254,486	46.1	268,358	48.6	29,007	5.3	551,851	100.0
	Silver birch	17,171	25.9	43,497	65.7	5,516	8.3	66,185	100.0
	Downy birch	11,572	8.4	99,257	72.3	26,400	19.2	137,229	100.0
	Other broadleaved	5,240	9.0	37,141	63.5	16,077	27.5	58,458	100.0
	Total	495,699	35.8	783,883	56.5	106,686	7.7	1,386,267	100.0
North Finland	Pine	72,772	20.2	257,480	71.5	29,991	8.3	360,244	100.0
	Norway spruce	25,142	24.0	69,305	66.1	10,462	10.0	104,909	100.0
	Silver birch	401	6.1	5,291	80.5	884	13.4	6,577	100.0
	Downy birch	796	0.8	64,690	68.3	29,254	30.9	94,740	100.0
	Other broadleaved	412	3.9	6,312	59.5	3,883	36.6	10,608	100.0
	Total	99,524	17.2	403,079	69.8	74,475	12.9	577,078	100.0
Whole country	Pine	280,002	30.0	593,110	63.6	59,677	6.4	932,789	100.0
	Norway spruce	279,628	42.6	337,663	51.4	39,469	6.0	656,760	100.0
	Silver birch	17,573	24.2	48,788	67.1	6,400	8.8	72,761	100.0
	Downy birch	12,368	5.3	163,947	70.7	55,654	24.0	231,969	100.0
	Other broadleaved	5,652	8.2	43,454	62.9	19,960	28.9	69,065	100.0
	Total	595,223	30.3	1,186,962	60.5	181,160	9.2	1,963,345	100.0

Table A.17 The number of stems and mean volume by tree species on forest land and poorly productive forest land in whole country

	Number of stems			Mean volume		
Tree species	stems/ha	%	Proportion of small trees[a] %	m³/ha	%	Proportion of small trees[a] %
Scots pine	801	23.9	28.5	43.4	47.8	0.3
Norway spruce	511	15.2	30.1	30.2	33.2	0.3
Silver birch	109	3.2	59.5	3.3	3.6	0.8
Downy birch	1,408	42.0	67.2	10.9	12.0	3.3
European aspen	95	2.8	70.9	1.4	1.6	1.9
Grey alder	141	4.2	56.1	1.0	1.1	4.5
Black alder	5	0.2	28.9	0.2	0.2	0.4
European mountain ash	238	7.1	90.2	0.2	0.2	31.7
Goat willow	28	0.8	49.7	0.3	0.3	3.2
Lodgepole pine	1	0.0	19.3	0.0	0.0	1.4
Larch	0	0.0	20.0	0.0	0.0	0.6
Fir	0	0.0	0.0	0.0	0.0	0.0
Common juniper	7	0.2	75.3	0.0	0.0	24.7
Other conifer	0	0.0	0.0	0.0	0.0	0.0
Bay willow	0	0.0	0.0	0.0	0.0	0.0
Wych elm	0	0.0	0.0	0.0	0.0	0.0
Small-leaved lime	0	0.0	0.0	0.0	0.0	0.0
European ash	0	0.0	38.0	0.0	0.0	1.7
Pedunculate oak	0	0.0	53.7	0.0	0.0	0.7
Bird cherry	8	0.2	71.9	0.0	0.0	17.1
Norway maple	0	0.0	54.5	0.0	0.0	3.9
Other broadleaved	2	0.1	87.1	0.0	0.0	14.7
Total	3,357	100.0	53.2	90.9	100.0	0.8

[a]Small trees are trees 0–2 cm at breast height

Table A.18 The diameter distribution of the growing stock by tree species on forest land and poorly productive forest land in whole country

Tree species	Diameter at breast height, cm																Total
	0–2		3–4		5–9		10–14		15–19		20–24		25–29		30–34		35–39		Over 39		
	stems/ha	%	stems/ha	%	stems/ha	%	stems/ha	%	stems/ha	%	stems/ha	%	stems/ha	%	stems/ha	%	stems/ha	%	stems/ha	%	stems/ha
Pine	233.9	28.9	125.2	15.5	203.8	25.2	124.3	15.4	67.2	8.3	30.0	3.7	13.9	1.7	6.6	0.8	2.8	0.4	1.4	0.2	809
Norway spruce	154.1	30.1	97.3	19.0	126.3	24.7	60.2	11.8	34.6	6.8	19.8	3.9	10.8	2.1	5.0	1.0	2.1	0.4	1.1	0.2	511
Silver birch	64.9	59.5	15.5	14.2	14.6	13.4	6.6	6.1	3.7	3.4	1.9	1.8	1.0	0.9	0.5	0.5	0.2	0.2	0.1	0.1	109
Downy birch	946.9	67.2	210.9	15.0	174.2	12.4	52.6	3.7	17.0	1.2	4.9	0.3	1.4	0.1	0.4	0.0	0.1	0.0	0.0	0.0	1,408
Aspen	67.6	70.9	11.9	12.4	9.7	10.2	3.2	3.4	1.4	1.5	0.7	0.8	0.4	0.4	0.2	0.2	0.1	0.1	0.1	0.1	95
Other broad-leaved	317.1	74.9	61.2	14.5	34.4	8.1	7.7	1.8	2.0	0.5	0.6	0.1	0.2	0.0	0.1	0.0	0.0	0.0	0.0	0.0	423
Total	1,784.5	53.2	521.9	15.5	563.0	16.8	254.7	7.6	125.9	3.8	58.0	1.7	27.6	0.8	12.7	0.4	5.4	0.2	2.8	0.1	3,357

Table A.19 The growing stock volume by tree species and by diameter classes on forest land and poorly productive forest land in whole country

Diameter at breast height, cm

Tree species	0–2 1,000 m³	%	3–4 1,000 m³	%	5–9 1,000 m³	%	10–14 1,000 m³	%	15–19 1,000 m³	%	20–24 1,000 m³	%	25–29 1,000 m³	%	30–34 1,000 m³	%	35–39 1,000 m³	%	Over 39 1,000 m³	%	Total 1,000 m³
Pine	3,073	0.3	8,125	0.8	68,950	6.9	161,049	16.1	208,914	20.9	185,554	18.6	148,658	14.9	106,754	10.7	63,399	6.3	45,163	4.5	999,640
Norway spruce	2,142	0.3	5,700	0.8	38,674	5.6	80,006	11.5	119,573	17.2	138,765	20.0	127,383	18.3	89,292	12.9	51,675	7.4	41,473	6.0	694,682
Silver birch	596	0.8	1,041	1.4	5,293	7.0	9,873	13.1	13,647	18.1	13,944	18.5	11,965	15.9	9,181	12.2	5,397	7.2	4,331	5.8	75,268
Downy birch	8,283	3.3	13,334	5.3	53,607	21.5	66,814	26.7	53,712	21.5	31,200	12.5	14,763	5.9	5,649	2.3	1,737	0.7	815	0.3	249,914
Aspen	616	1.9	759	2.3	3,568	11.0	5,006	15.4	5,426	16.7	5,059	15.6	4,126	12.7	3,114	9.6	2,085	6.4	2,758	8.5	32,517
Other broad-leaved	2,909	7.4	3,916	10.0	10,560	27.0	9,564	24.4	5,971	15.2	3,205	8.2	1,500	3.8	822	2.1	286	0.7	439	1.1	39,172
Total	17,620	0.8	32,875	1.6	180,652	8.6	332,312	15.9	407,243	19.5	377,727	18.1	308,394	14.7	214,812	10.3	124,578	6.0	94,980	4.5	2,091,193

Appendix

Table A.20 The number of stems and mean volume of the saw-timber sized stock by diameter classes and tree species

	Number of stems								Mean volume							
	Diameter at breast height, cm								Diameter at breast height, cm							
Tree species	15–19	20–24	25–29	30–34	35–39	40–44	Over 44	Total	15–19	20–24	25–29	30–34	35–39	40–44	Over 44	Total
	stems/ha								m³/ha							

(a) The number of stems and mean volume of the saw-timber sized stock on forest land

South Finland

Pine	5.1	19.9	14.5	7.9	3.4	1.1	0.4	52.3	1.1	6.2	7.4	6.0	3.6	1.6	0.7	26.6
Norway spruce	4.0	21.9	16.1	8.1	3.5	1.3	0.6	55.5	1.0	7.4	8.9	6.7	3.9	2.0	1.2	31.0
Birch	0.1	3.9	2.8	1.2	0.4	0.1	0.1	8.6	0.0	1.4	1.5	1.0	0.5	0.2	0.1	4.7
Other broadleaved	0.0	0.5	0.4	0.2	0.1	0.0	0.0	1.3	0.0	0.2	0.2	0.2	0.1	0.1	0.1	0.8
Total	9.2	46.2	33.8	17.4	7.4	2.7	1.0	117.6	2.1	15.2	18.0	13.8	8.1	3.9	2.0	63.0

North Finland

Pine	4.8	16.1	8.8	3.7	1.4	0.5	0.1	35.3	0.9	4.4	3.9	2.5	1.3	0.5	0.2	13.8
Norway spruce	0.6	5.7	3.6	1.6	0.7	0.2	0.1	12.6	0.1	1.6	1.6	1.0	0.6	0.2	0.2	5.3
Birch	0.0	0.5	0.3	0.0	0.0	0.0	0.0	0.9	0.0	0.2	0.1	0.0	0.0	0.0	0.0	0.3
Other broadleaved	0.0	0.1	0.1	0.0	0.0	0.0	0.0	0.2	0.0	0.0	0.0	0.0	0.0	0.0	0.0	0.1
Total	5.4	22.5	12.7	5.4	2.1	0.7	0.3	48.9	1.1	6.2	5.6	3.5	1.9	0.8	0.4	19.5

Whole country

Pine	4.9	18.2	11.9	6.0	2.5	0.8	0.3	44.6	1.0	5.4	5.8	4.4	2.5	1.1	0.5	20.8
Norway spruce	2.5	14.6	10.5	5.2	2.2	0.8	0.4	36.2	0.6	4.8	5.6	4.1	2.4	1.2	0.7	19.4
Birch	0.0	2.4	1.6	0.7	0.2	0.1	0.0	5.1	0.0	0.8	0.9	0.5	0.3	0.1	0.1	2.7
Other broadleaved	0.0	0.3	0.3	0.1	0.1	0.0	0.0	0.8	0.0	0.1	0.1	0.1	0.1	0.0	0.0	0.5
Total	7.5	35.5	24.3	12.0	5.0	1.8	0.7	86.7	1.6	11.2	12.4	9.2	5.3	2.5	1.3	43.4

(continued)

Table A.20 (continued)

(b) The number of stems and mean volume of the saw-timber sized stock on forest land available for wood production

Tree species	Number of stems stems/ha								Mean volume m³/ha							
	Diameter at breast height, cm								Diameter at breast height, cm							
	15–19	20–24	25–29	30–34	35–39	40–44	Over 44	Total	15–19	20–24	25–29	30–34	35–39	40–44	Over 44	Total
South Finland
Pine	5.1	19.8	14.4	7.8	3.3	1.1	0.3	51.9	1.1	6.2	7.3	6.0	3.5	1.6	0.6	26.3
Norway spruce	4.0	22.0	16.2	8.1	3.4	1.3	0.6	55.6	1.0	7.5	8.9	6.6	3.9	2.0	1.2	31.0
Birch	0.1	3.9	2.7	1.2	0.4	0.1	0.1	8.5	0.0	1.4	1.5	0.9	0.5	0.2	0.1	4.6
Other broadleaved	0.0	0.5	0.4	0.2	0.1	0.0	0.0	1.3	0.0	0.2	0.2	0.2	0.1	0.1	0.0	0.8
Total	9.1	46.1	33.8	17.4	7.3	2.6	1.0	117.3	2.1	15.2	18.0	13.7	8.0	3.8	2.0	62.8
North Finland
Pine	4.9	16.5	8.5	3.3	1.1	0.4	0.1	34.8	1.0	4.5	3.8	2.2	1.1	0.4	0.2	13.1
Norway spruce	0.6	5.1	2.9	1.2	0.4	0.1	0.1	10.5	0.1	1.4	1.3	0.8	0.4	0.1	0.1	4.3
Birch	0.0	0.5	0.2	0.0	0.0	0.0	0.0	0.8	0.0	0.2	0.1	0.0	0.0	0.0	0.0	0.3
Other broadleaved	0.0	0.1	0.1	0.0	0.0	0.0	0.0	0.2	0.0	0.0	0.0	0.0	0.0	0.0	0.0	0.1
Total	5.5	22.2	11.7	4.6	1.6	0.5	0.2	46.3	1.1	6.1	5.2	3.0	1.5	0.6	0.3	17.8
Whole country
Pine	5.0	18.4	11.9	5.9	2.4	0.8	0.2	44.6	1.0	5.5	5.8	4.4	2.5	1.1	0.4	20.7
Norway spruce	2.6	14.8	10.5	5.2	2.2	0.8	0.4	36.4	0.6	4.9	5.7	4.2	2.4	1.2	0.7	19.6
Birch	0.0	2.5	1.7	0.7	0.3	0.1	0.0	5.3	0.0	0.9	0.9	0.6	0.3	0.1	0.1	2.8
Other broadleaved	0.0	0.3	0.3	0.1	0.1	0.0	0.0	0.9	0.0	0.1	0.1	0.1	0.1	0.0	0.0	0.5
Total	7.6	36.0	24.4	11.9	4.9	1.7	0.6	87.1	1.7	11.4	12.6	9.2	5.2	2.4	1.2	43.6

Table A.21 The annual increment of the growing stock on forest land and on poorly productive forest land

	Forest land			Poorly productive forest land			Total		
	m³/ha/year	1,000 m³/year	%	m³/ha/year	1,000 m³/year	%	m³/ha/year	1,000 m³/year	%

(a) The annual increment by tree species groups on forest land and on poorly productive forest land

South Finland

Pine	2.1	23,084	38.5	0.6	243	74.1	2.0	23,328	38.7
Norway spruce	2.1	22,902	38.2	0.0	11	3.3	2.0	22,913	38.0
Birch	0.9	10,107	16.9	0.1	58	17.6	0.9	10,164	16.9
Other broadleaved	0.3	3,851	6.4	0.0	16	5.0	0.3	3,867	6.4
Total	5.4	59,944	100.0	0.8	329	100.0	5.2	60,273	100.0

North Finland

Pine	1.7	15,521	61.1	0.3	714	60.8	1.4	16,235	61.1
Norway spruce	0.5	4,287	16.9	0.1	136	11.6	0.4	4,423	16.7
Birch	0.5	4,998	19.7	0.1	309	26.3	0.5	5,307	20.0
Other broadleaved	0.1	578	2.3	0.0	16	1.3	0.1	594	2.2
Total	2.8	25,384	100.0	0.5	1,174	100.0	2.3	26,559	100.0

Whole country

Pine	1.9	38,605	45.2	0.4	957	63.7	1.7	39,562	45.6
Norway spruce	1.3	27,189	31.9	0.1	147	9.8	1.2	27,336	31.5
Birch	0.7	15,104	17.7	0.1	367	24.4	0.7	15,471	17.8
Other broadleaved	0.2	4,430	5.2	0.0	32	2.1	0.2	4,461	5.1
Total	4.2	85,328	100.0	0.6	1,503	100.0	3.8	86,831	100.0

(b) The annual increment by principal site classes on forest land and on poorly productive forest land

South Finland

Mineral soil

| Undrained | 5.6 | 43,678 | | 0.9 | 147 | | 5.5 | 43,825 | |
| Drained | 5.2 | 3,870 | | – | – | | 5.2 | 3,870 | |

(continued)

Table A.21 (continued)

	Forest land			Poorly productive forest land			Total		
	m³/ha/year	1,000 m³/year	%	m³/ha/year	1,000 m³/year	%	m³/ha/year	1,000 m³/year	%
Pine	2.1	17,747	37.3	0.7	110	74.8	2.1	17,857	37.4
Norway spruce	2.3	19,329	40.7	0.1	8	5.7	2.2	19,337	40.5
Birch	0.8	6,955	14.6	0.1	18	12.5	0.8	6,973	14.6
Other broadleaved	0.4	3,518	7.4	0.1	10	7.0	0.4	3,528	7.4
Total	5.6	47,548	100.0	0.9	147	100.0	5.5	47,695	100.0
Peatland									
Undrained	4.3	1,405		0.8	98		3.3	1,504	
Drained	4.7	10,990		0.6	84		4.5	11,074	
Pine	2.0	5,337	43.1	0.5	134	73.5	1.9	5,471	43.5
Norway spruce	1.3	3,573	28.8	0.0	3	1.4	1.2	3,576	28.4
Birch	1.2	3,152	25.4	0.1	40	21.8	1.1	3,192	25.4
Other broadleaved	0.1	333	2.7	0.0	6	3.3	0.1	339	2.7
Total	4.6	12,395	100.0	0.7	182	100.0	4.3	12,578	100.0
North Finland									
Mineral soil									
Undrained	2.5	15,797		0.5	390		2.3	16,187	
Drained	3.6	1,737		–	–		3.6	1,737	
Pine	1.7	11,390	65.0	0.2	150	38.4	1.6	11,539	64.4
Norway spruce	0.4	2,970	16.9	0.1	42	10.7	0.4	3,012	16.8
Birch	0.4	2,722	15.5	0.3	189	48.6	0.4	2,911	16.2
Other broadleaved	0.1	453	2.6	0.0	9	2.4	0.1	462	2.6
Total	2.6	17,535	100.0	0.5	390	100.0	2.4	17,925	100.0

Appendix

Peatland										
Undrained	2.1	1,091		0.5	527			1.0	1,618	
Drained	3.5	6,759		0.7	257			3.0	7,016	
Pine	1.7	4,131	52.6	0.4	564	71.9		1.2	4,695	54.4
Norway spruce	0.5	1,317	16.8	0.1	95	12.0		0.4	1,412	16.4
Birch	0.9	2,276	29.0	0.1	120	15.2		0.6	2,395	27.7
Other broadleaved	0.1	126	1.6	0.0	6	0.8		0.0	132	1.5
Total	3.2	7,850	100.0	0.5	784	100.0		2.2	8,634	100.0
Whole country										
Mineral soil										
Undrained	4.3	59,475		0.6	536			4.0	60,012	
Drained	4.6	5,608		–	–			4.6	5,608	
Pine	1.9	29,137	44.8	0.3	259	48.3		1.8	29,396	44.8
Norway spruce	1.5	22,299	34.3	0.1	50	9.3		1.4	22,349	34.1
Birch	0.6	9,677	14.9	0.2	208	38.7		0.6	9,884	15.1
Other broadleaved	0.3	3,970	6.1	0.0	20	3.7		0.2	3,990	6.1
Total	4.3	65,083	100.0	0.6	536	100.0		4.1	65,619	100.0
Peatland										
Undrained	3.0	2,496		0.5	625			1.5	3,122	
Drained	4.1	17,749		0.7	341			3.7	18,090	
Pine	1.8	9,468	46.8	0.4	698	72.2		1.5	10,166	47.9
Norway spruce	0.9	4,890	24.2	0.1	97	10.1		0.7	4,988	23.5
Birch	1.1	5,428	26.8	0.1	159	16.5		0.8	5,587	26.3
Other broadleaved	0.1	459	2.3	0.0	12	1.3		0.1	471	2.2
Total	3.9	20,245	100.0	0.5	966	100.0		3.1	21,212	100.0

(continued)

Table A.21 (continued)

(c) The annual increment by tree species groups on forest land and on poorly productive forest land available for wood production

South Finland

	Forest land			Poorly productive forest land			Total		
	m³/ha/year	1,000 m³/year	%	m³/ha/year	1,000 m³/year	%	m³/ha/year	1,000 m³/year	%
Pine	2.1	22,771	38.4	0.6	225	73.9	2.0	22,996	38.6
Norway spruce	2.1	22,712	38.3	0.0	11	3.5	2.0	22,723	38.2
Birch	0.9	9,988	16.9	0.1	54	17.6	0.9	10,041	16.9
Other broadleaved	0.3	3,775	6.4	0.0	15	5.0	0.3	3,790	6.4
Total	5.4	59,246	100.0	0.8	304	100.0	5.2	59,550	100.0

North Finland

	m³/ha/year	1,000 m³/year	%	m³/ha/year	1,000 m³/year	%	m³/ha/year	1,000 m³/year	%
Pine	1.8	14,653	61.8	0.4	576	69.3	1.6	15,229	62.0
Norway spruce	0.5	3,785	16.0	0.1	85	10.2	0.4	3,870	15.8
Birch	0.6	4,746	20.0	0.1	159	19.1	0.5	4,905	20.0
Other broadleaved	0.1	545	2.3	0.0	12	1.4	0.1	557	2.3
Total	2.9	23,729	100.0	0.5	831	100.0	2.5	24,561	100.0

Whole country

	m³/ha/year	1,000 m³/year	%	m³/ha/year	1,000 m³/year	%	m³/ha/year	1,000 m³/year	%
Pine	1.9	37,424	45.1	0.4	801	70.5	1.8	38,225	45.4
Norway spruce	1.4	26,497	31.9	0.1	96	8.4	1.3	26,593	31.6
Birch	0.8	14,734	17.8	0.1	213	18.7	0.7	14,946	17.8
Other broadleaved	0.2	4,320	5.2	0.0	27	2.4	0.2	4,347	5.2
Total	4.3	82,975	100.0	0.6	1,136	100.0	4.0	84,111	100.0

Table A.22 The proportional areas of silvicultural quality classes by development classes on forest land available for wood production

Stand quality	Temporarily unstocked stands	Young seedling stands	Advanced seedling stands	Young thinning stands	Advanced thinning stands	Mature stands	Shelter tree stands	Seed tree stands	Total
	% of area								
South Finland									
Good	63.2	60.2	41.3	38.2	48.1	49.6	31.6	59.9	45.1
Satisfactory	5.2	25.4	37.0	42.0	40.3	33.2	14.8	13.3	37.4
Passable	16.5	10.6	16.4	14.9	9.6	9.2	18.3	12.5	12.6
Low-yielding	15.0	3.8	5.3	4.8	2.1	7.9	35.3	14.2	4.9
Total	100.0	100.0	100.0	100.0	100.0	100.0	100.0	100.0	100.0
Area of the development class, km^2	1,620	7,384	16,499	37,128	29,794	16,232	427	1,295	110,379
North Finland									
Good	61.6	44.4	20.9	15.7	22.6	10.4	3.1	28.9	19.7
Satisfactory	5.7	35.5	40.9	46.9	47.6	25.2	10.8	17.6	41.6
Passable	12.1	14.2	31.0	27.8	21.9	22.5	14.8	20.5	25.4
Low-yielding	20.7	5.9	7.1	9.5	7.9	41.9	71.4	33.0	13.3
Total	100.0	100.0	100.0	100.0	100.0	100.0	100.0	100.0	100.0
Area of the development class, km^2	992	5,330	13,032	36,511	14,319	10,057	232	1,188	81,662
Whole country									
Good	62.6	53.6	32.3	27.1	39.8	34.6	21.6	45.1	34.3
Satisfactory	5.4	29.6	38.7	44.4	42.7	30.1	13.4	15.4	39.2
Passable	14.8	12.1	22.9	21.3	13.6	14.3	17.1	16.3	18.0
Low-yielding	17.2	4.7	6.1	7.1	4.0	21.0	48.0	23.2	8.5
Total	100.0	100.0	100.0	100.0	100.0	100.0	100.0	100.0	100.0
Area of the development class, km^2	2,612	12,714	29,531	73,640	44,113	2,689	660	2,482	192,041

Table A.23 The areas of seedling and young thinning stands by establishment method on forest land

| | Natural regeneration | | Artificial regeneration | | | | Total | |
| | | | Successful | | Failed | | | |
	km²	%	km²	%	km²	%	km²	%
South Finland								
Old forest land[a]	31,968	54.5	25,408	43.3	1,319	2.2	58,695	100.0
New forest land	970	34.0	1,691	59.2	196	6.9	2,857	100.0
Young stands total	32,938	53.5	27,099	44.0	1,515	2.5	61,552	100.0
North Finland								
Old forest land[a]	38,186	68.7	15,875	28.6	1,490	2.7	55,550	100.0
New forest land	926	74.5	257	20.7	60	4.8	1,243	100.0
Young stands total	39,112	68.9	16,132	28.4	1,549	2.7	56,793	100.0
Whole country								
Old forest land[a]	70,154	61.4	41,283	36.1	2,809	2.5	114,246	100.0
New forest land	1,897	46.3	1,948	47.5	256	6.2	4,100	100.0
Young stands total	72,051	60.9	43,231	36.5	3,064	2.6	118,346	100.0

[a] Forest land was old if it had been forest for the last 30 years, otherwise it was new forest land

Table A.24 The proportional areas of the seedling stands by dominant tree species and by the number of the seedlings capable for future development on forest land

Dominant tree species	Number of seedlings/ha								Total
	1–450	451–950	951–1,450	1,451–1,950	1,951–2,950	2,951–3,950	3,951–4,950	over 4,950	
	Proportion of area, %								
South Finland									
Young seedling stands									
Pine	0.2	1.0	9.7	17.3	33.1	17.7	9.9	11.1	100.0
Norway spruce	0.8	1.6	10.9	22.2	39.6	15.0	5.5	4.2	100.0
Broadleaved	2.2	10.5	14.9	21.8	23.4	10.7	6.9	9.8	100.0
Advanced seedling stands									
Pine	0.2	2.2	13.7	22.5	42.3	13.8	3.9	1.4	100.0
Norway spruce	0.4	2.2	15.4	22.2	39.8	14.5	3.5	2.1	100.0
Broadleaved	8.4	10.9	14.1	19.6	29.6	10.2	3.9	3.2	100.0
North Finland									
Young seedling stands									
Pine	0.8	4.7	15.1	21.0	30.0	14.5	8.7	5.1	100.0
Norway spruce	2.4	3.9	17.4	21.8	33.3	14.3	3.9	3.1	100.0
Broadleaved	15.5	40.8	9.3	12.5	15.7	3.1	0.0	3.1	100.0
Advanced seedling stands									
Pine	1.1	12.3	30.8	25.7	23.4	5.2	1.1	0.3	100.0
Norway spruce	3.2	8.1	22.7	23.4	31.2	10.8	0.5	0.0	100.0
Broadleaved	25.5	19.7	14.7	13.9	17.9	4.5	2.5	1.2	100.0

(continued)

Table A.24 (continued)

Dominant tree species	Number of seedlings/ha							Total
	1–450	451–950	951–1,450	1,451–1,950	1,951–2,950	2,951–3,950	3,951–4,950	over 4,950	
	Proportion of area, %							
Whole country									
Young seedling stands									
Pine	0.5	3.0	12.6	19.3	31.5	16.0	9.2	7.9	100.0
Norway spruce	1.2	2.2	12.5	22.1	38.1	14.8	5.1	3.9	100.0
Broadleaved	3.9	14.3	14.2	20.7	22.4	9.7	6.0	8.9	100.0
Advanced seedling stands									
Pine	0.7	7.5	22.7	24.2	32.4	9.3	2.4	0.8	100.0
Norway spruce	1.2	3.9	17.5	22.5	37.4	13.5	2.6	1.5	100.0
Broadleaved	12.1	12.9	14.3	18.3	27.1	8.9	3.6	2.8	100.0

Appendix   235

Table A.25 The areas of the most recent cuttings in the previous 10-year period by cutting time and method on forest land and poorly productive forest land

Cutting time[a]	Cutting method										Total	Proportion of forest land area
	1	2	3	4	6	7	8	9	10			
	km²											%
South Finland												
Previous logging year	1,018	318	991	983	319	558	294	17	26		4,524	4.1
Previous logging years 2–5	4,254	1,149	4,497	5,701	856	3,341	1,984	48	142		21,973	19.7
Previous logging years 6–10	5,436	668	2,209	4,275	363	2,840	1,177	71	58		17,097	15.3
Previous 10-year period total	10,708	2,136	7,698	10,958	1,538	6,739	3,456	136	225		43,594	39.0
North Finland												
Previous logging year	513	229	465	321	141	269	99	7	7		2,051	2.2
Previous logging years 2–5	1,765	697	1,955	1,231	561	1,617	601	11	138		8,576	9.4
Previous logging years 6–10	2,429	606	1,335	999	364	1,791	634	21	68		8,246	9.0
Previous 10-year period total	4,707	1,532	3,756	2,551	1,065	3,676	1,334	39	213		18,873	20.6
Whole country												
Previous logging year	1,531	548	1,457	1,304	460	827	393	24	33		6,576	3.2
Previous logging years 2–5	6,020	1,846	6,453	6,932	1,417	4,958	2,585	59	280		30,548	15.0
Previous logging years 6–10	7,865	1,274	3,544	5,274	728	4,631	1,812	92	125		25,343	12.5
Previous 10-year period total	15,416	3,668	11,453	13,509	2,604	10,415	4,790	175	438		62,467	30.7

Cutting methods: *1* Tending of seedling stands, *2* Removal of standards, *3* First thinning, *4* Other thinning, *6* Special cuttings, e.g. post-damage sanitary cutting, opening of road or ditch lines and slight removal of standard-like trees, *7* Regeneration cutting for artificial regeneration, *8* Regeneration cutting for natural regeneration, *9* Cutting of nurse crop, *10* Selection cutting

[a] The logging year begins on 1st of June

Table A.26 The areas of the proposed cuttings for the coming 10-year period by cutting method on forest land available for wood production

Timing of cutting	Cutting method									Total	Proportion of forest land area available for wood production
	1	2	3	4	6	7	8	9			%
	km²										
South Finland											
Delayed	2,773	1,100	3,190	2,663	14	1,434	615	263		12,053	10.9
First 5-year period, not delayed	6,947	2,198	7,779	7,013	29	8,330	3,854	186		36,336	32.9
Second 5-year period	5,007	719	8,979	5,382	3	2,980	1,320	6		24,396	22.1
Total	14,726	4,017	19,948	15,059	46	12,744	5,789	455		72,785	65.9
North Finland											
Delayed	3,165	1,239	2,137	822	0	4,768	1,333	77		13,541	16.6
First 5-year period, not delayed	3,380	690	3,707	2,032	14	4,727	1,551	21		16,122	19.7
Second 5-year period	2,969	501	5,429	1,866	0	1,225	427	0		12,418	15.2
Total	9,513	2,431	11,272	4,720	14	10,721	3,311	99		42,081	51.5
Whole country											
Delayed	5,938	2,339	5,327	3,485	14	6,202	1,948	341		25,594	13.3
First 5-year period, not delayed	10,326	2,889	11,485	9,045	43	13,057	5,405	208		52,458	27.3
Second 5-year period	7,975	1,220	14,408	7,249	3	4,206	1,748	6		36,814	19.2
Total	24,240	6,448	31,220	19,779	60	23,465	9,100	554		114,865	59.8

Cutting methods: *1* Tending of seedling stands, *2* Removal of standards, *3* First thinning, *4* Other thinnings, *6* Special cuttings, mostly post-damage sanitary cuttings, *7* Regeneration cutting for artificial regeneration, *8* Regeneration cutting for natural regeneration, *9* Cutting of nurse crop

Table A.27 The areas of the most recent cutting by cutting time on forest land and on poorly productive forest land

Cutting time[a]	Forest land km²	Proportion of land class area %	Poorly productive forest land km²	Proportion of land class area %
South Finland				
Current logging year	935	0.8	6	0.1
Previous logging year	4,524	4.1	9	0.2
Previous logging years 2–5	21,973	19.7	105	2.4
Previous logging years 6–10	17,097	15.3	104	2.4
Previous logging years 11–30	49,429	44.3	555	13.0
More than 30 logging years or no logging	17,712	15.9	3,502	81.8
Total	111,671	100.0	4,280	100.0
North Finland				
Current logging year	403	0.4	11	0.0
Previous logging year	2,051	2.2	25	0.1
Previous logging years 2–5	8,576	9.4	159	0.7
Previous logging years 6–10	8,246	9.0	174	0.8
Previous logging years 11–30	33,847	36.9	2,684	12.0
More than 30 logging years or no logging	38,585	42.1	19,370	86.4
Total	91,707	100.0	22,422	100.0
Whole country				
Current logging year	1,337	0.7	16	0.1
Previous logging year	6,576	3.2	33	0.1
Previous logging years 2–5	30,548	15.0	264	1.0
Previous logging years 6–10	25,343	12.5	278	1.0
Previous logging years 11–30	83,276	40.9	3,238	12.1
More than 30 logging years or no logging	56,297	27.7	22,872	85.7
Total	203,378	100.0	26,701	100.0

[a] The logging year begins on 1st of June

Table A.28 The areas of the accomplished silvicultural measures during the previous 10-year period on forest land

Timing of measure	Planting or seeding km²	Proportion of forest land area %	Complementary planting or seeding km²	Proportion of forest land area %	Pruning km²	Proportion of forest land area %
South Finland						
Previous year	748	0.7	42	0.0	104	0.1
Previous years 2–5	2,892	2.6	102	0.1	753	0.7
Previous years 6–10	3,234	2.9	81	0.1	502	0.4
10-year period total	6,873	6.2	225	0.2	1,359	1.2
North Finland						
Previous year	272	0.3	4	0.0	14	0.0
Previous years 2–5	1,438	1.6	25	0.0	71	0.1
Previous years 6–10	2,209	2.4	74	0.1	145	0.2
10-year period total	3,919	4.3	103	0.1	230	0.3
Whole country						
Previous year	1,020	0.5	46	0.0	118	0.1
Previous years 2–5	4,330	2.1	127	0.1	824	0.4
Previous years 6–10	5,442	2.7	155	0.1	646	0.3
10-year period total	10,792	5.3	328	0.2	1,589	0.8

Table A.29 The areas of the silvicultural measures proposed at the current stage of the forest stand for the coming 10-year period on forest land available for wood production

Proposed measure	Area km²	Proportion of forest land area available for wood production %
South Finland		
Planting or seeding	2,196	2.0
Complementary planting or seeding	368	0.3
Weeding of seedling stand	108	0.1
Cleaning of regeneration area	165	0.1
Cleaning and planting or seeding	505	0.5
North Finland		
Planting or seeding	1,788	2.2
Complementary planting or seeding	195	0.2
Weeding of seedling stand	7	0.0
Cleaning of regeneration area	105	0.1
Cleaning and planting or seeding	483	0.6
Whole country		
Planting or seeding	3,984	2.1
Complementary planting or seeding	563	0.3
Weeding of seedling stand	115	0.1
Cleaning of regeneration area	269	0.1
Cleaning and planting or seeding	988	0.5

Appendix

Table A.30 The areas of the accomplished soil preparation during the previous 30-year period on forest land

Timing of soil preparation	Light soil preparation[a] km²	Proportion of forest land area %	Ploughing km²	Proportion of forest land area %	Mounding km²	Proportion of forest land area %	Prescribed burning km²	Proportion of forest land area %	Total km²	Proportion of forest land area %
South Finland										
Previous year	936	0.8	8	0.0	128	0.1	11	0.0	1,083	1.0
Previous years 2–5	3,195	2.9	46	0.0	442	0.4	48	0.0	3,731	3.3
Previous years 6–10	2,935	2.6	141	0.1	465	0.4	28	0.0	3,570	3.2
Previous 10-year period total	7,066	6.3	195	0.2	1,035	0.9	88	0.1	8,384	7.5
Previous years 11–30	8,051	7.2	1,132	1.0	496	0.4	123	0.1	9,801	8.8
North Finland										
Previous year	350	0.4	61	0.1	105	0.1	52	0.1	568	0.6
Previous years 2–5	1,184	1.3	211	0.2	282	0.3	18	0.0	1,694	1.8
Previous years 6–10	1,307	1.4	579	0.6	285	0.3	35	0.0	2,206	2.4
Previous 10-year period total	2,841	3.1	852	0.9	672	0.7	105	0.1	4,469	4.9
Previous years 11–30	3,772	4.1	5,963	6.5	335	0.4	209	0.2	10,279	11.2
Whole country										
Previous year	1,286	0.6	69	0.0	233	0.1	63	0.0	1,652	0.8
Previous years 2–5	4,379	2.2	257	0.1	723	0.4	66	0.0	5,425	2.7
Previous years 6–10	4,242	2.1	720	0.4	751	0.4	64	0.0	5,776	2.8
Previous 10-year period total	9,907	4.9	1,046	0.5	1,707	0.8	193	0.1	12,853	6.3
Previous years 11–30	11,822	5.8	7,095	3.5	830	0.4	332	0.2	20,080	9.9

[a]Harrowing or scarification

Table A.31 The areas of the soil preparation proposed at the current stage of the forest stand, and after the proposed regeneration cutting for the coming 10-year period on forest land available for wood prodiction

Timing of proposed soil preparation	Light soil preparation[a] km²	Proportion of forest land area available for wood production %	Ploughing km²	Proportion of forest land area available for wood production %	Mounding km²	Proportion of forest land area available for wood production %	Total km²	Proportion of forest land area available for wood production %
South Finland								
At the current stage	2,034	1.8	17	0.0	534	0.5	2,586	2.3
After regeneration cutting	13,803	12.5	121	0.1	2,442	2.2	16,366	14.8
North Finland								
At the current stage	1,673	2.0	213	0.3	536	0.7	2,422	3.0
After regeneration cutting	8,020	9.8	2,113	2.6	3,285	4.0	13,417	16.4
Whole country								
At the current stage	3,707	1.9	230	0.1	1,071	0.6	5,008	2.6
After regeneration cutting	21,823	11.4	2,234	1.2	5,727	3.0	29,783	15.5

[a] Harrowing or scarification

Appendix

Table A.32 The areas of accomplished forest drainages and other water balance related measures during the previous 10-year period and accomplished forest drainage activities in 11–30 years preceeding the inventory on forest land, and on peatland also on poorly productive forest land and unproductive land

	Mineral soil	Peatland				
	Forest land	Forest land	Poorly productive forest land	Unproductive land	Total	Total
Measure and timing	km²					
South Finland						
Forest drainage, previous 10 years						
First-time drainage	1,239	909	106	28	1,043	2,282
Cleaning of ditches	309	2,153	89	6	2,247	2,556
Complementary drainage[a]	187	1,323	31	3	1,357	1,543
Forest drainages total	1,734	4,384	226	37	4,647	6,381
Other measures, previous 10 years						
Non-forest drainage	46	31	3	6	40	86
Blocking of ditches	0	11	3	0	14	14
11–30 years old drainage	4,164	12,388	798	114	13,300	17,463
North Finland						
Forest drainage, previous 10 years						
First-time drainage	398	486	162	21	670	1,068
Cleaning of ditches	173	1,269	84	4	1,357	1,530
Complementary drainage[a]	102	928	53	0	981	1,083
Forest drainages total	673	2,684	300	25	3,008	3,681
Other measures, previous 10 years						
Non-forest drainage	25	32	21	25	77	102
Blocking of ditches	0	0	0	0	0	0
11–30 years old drainage	2,482	10,161	2,208	304	12,674	15,156
Whole country						
Forest drainage, previous 10 years						
First-time drainage	1,637	1,395	269	49	1,713	3,350
Cleaning of ditches	481	3,422	173	9	3,604	4,085
Complementary drainage[a]	289	2,251	84	3	2,338	2,626
Forest drainages total	2,407	7,068	526	61	7,655	10,062
Other measures, previous 10 years						
Non-forest drainage	71	63	24	30	117	188
Blocking of ditches	0	11	3	0	14	14
11–30 years old drainage	6,646	22,549	3,006	418	25,973	32,619

[a] Besides the complementary drainage, cleaning of ditches could also have been done on these areas

Table A.33 The areas of proposed forest drainage activities on forest land and poorly productive forest land available for wood production

	Mineral soil	Peatland			Total
	Forest land	Forest land	Poorly productive forest land	Total	
	km²				
South Finland					
First-time drainage[a]	1,390	2,554	171	2,724	4,114
Cleaning of ditches	353	6,378	17	6,395	6,748
Complementary drainage[b]	196	2,737	54	2,791	2,987
Total	1,939	11,668	242	11,910	13,849
North Finland					
First-time drainage[a]	1,716	2,909	679	3,588	5,303
Cleaning of ditches	138	4,225	56	4,281	4,419
Complementary drainage[b]	304	2,972	130	3,103	3,407
Total	2,157	10,106	865	10,972	13,129
Whole country					
First-time drainage[a]	3,105	5,463	849	6,312	9,417
Cleaning of ditches	491	10,602	74	10,676	11,167
Complementary drainage[b]	500	5,709	185	5,894	6,394
Total	4,096	21,775	1,108	22,882	26,978

[a] On peatland first-time drainage represents the area suitable for wood production after drainage
[b] Besides the complementary drainage, these areas may also require cleaning of ditches

Appendix 243

Table A.34 The areas of damaged forests by causal agents and by degree of the damage on forest land in whole country

Causal agent	Degree of damage								Total		Proportion of forest land area
	Slight		Moderate		Severe		Complete				
	km²	%	km²	%	km²	%	km²	%	km²	%	%
Abiotic total	7,068	15.7	8,068	20.6	2,683	26.6	234	32.1	18,053		8.9
Wind	1,464	3.2	644	1.6	229	2.3	30	4.1	2,368		1.2
Snow	3,541	7.8	4,237	10.8	1,470	14.6	77	10.5	9,325		4.6
Frost	337	0.7	336	0.9	43	0.4	6	0.8	722		0.4
Other climatic factors	220	0.5	232	0.6	68	0.7	3	0.4	523		0.3
Fire	40	0.1	95	0.2	144	1.4	30	4.2	309		0.2
Soil factors	1,466	3.2	2,524	6.4	728	7.2	89	12.2	4,807		2.4
Anthropogenic total	2,916	6.5	2,065	5.3	169	1.7	20	2.7	5,170		2.5
Logging	1,128	2.5	280	0.7	14	0.1	0	0.0	1,421		0.7
Air pollution	9	0.0	6	0.0	4	0.0	0	0.0	18		0.0
Other human activity	1,780	3.9	1,779	4.5	152	1.5	20	2.7	3,731		1.8
Animals total	4,607	10.2	3,354	8.6	831	8.2	125	17.2	8,918		4.4
Voles	40	0.1	28	0.1	20	0.2	0	0.0	88		0.0
Moose, deer or reindeer	2,848	6.3	2,828	7.2	760	7.5	94	13.0	6,530		3.2
Other vertebrates	104	0.2	63	0.2	28	0.3	31	4.3	226		0.1
Pine shoot beetles	405	0.9	111	0.3	0	0.0	0	0.0	515		0.3
Pine weevil	3	0.0	13	0.0	3	0.0	0	0.0	18		0.0
Pine sawflys	239	0.5	28	0.1	0	0.0	0	0.0	266		0.1
Common pine sawfly	616	1.4	63	0.2	3	0.0	0	0.0	681		0.3
European pine sawfly	187	0.4	55	0.1	0	0.0	0	0.0	242		0.1
Other defoliators	14	0.0	6	0.0	0	0.0	0	0.0	20		0.0
Spruce bark beetles	3	0.0	4	0.0	0	0.0	0	0.0	7		0.0
Other identified insect	37	0.1	60	0.2	11	0.1	0	0.0	107		0.1
Unidentified insect	112	0.2	97	0.2	7	0.1	0	0.0	216		0.1

(continued)

Table A.34 (continued)

Causal agent	Degree of damage								Total		Proportion of forest land area
	Slight		Moderate		Severe		Complete				
	km²	%	km²	%	km²	%	km²	%	km²		%
Fungi total	9,312	20.6	7,803	19.9	2,527	25.0	95	13.1	19,737		9.7
Annosum root rot	835	1.8	1,098	2.8	178	1.8	17	2.3	2,128		1.0
Other decay fungi	1,394	3.1	3,220	8.2	1,159	11.5	38	5.2	5,812		2.9
Scleroderris canker	3,022	6.7	874	2.2	93	0.9	9	1.2	3,998		2.0
Pine twisting rust	665	1.5	672	1.7	301	3.0	24	3.3	1,663		0.8
Blister rust	2,139	4.7	1,299	3.3	571	5.7	0	0.0	4,009		2.0
Other rust fungi	215	0.5	16	0.0	0	0.0	0	0.0	230		0.1
Needle cast fungi	710	1.6	388	1.0	120	1.2	0	0.0	1,218		0.6
Other identified fungi	92	0.2	65	0.2	3	0.0	0	0.0	160		0.1
Unidentified fungi	239	0.5	171	0.4	101	1.0	7	1.0	518		0.3
Competition between plants	1,515	3.4	1,295	3.3	136	1.3	14	2.0	2,961		1.5
Unknown	19,733	43.7	16,576	42.3	3,749	37.1	239	32.9	40,298		19.8
Total	45,152	100.0	39,161	100.0	10,096	100.0	728	100.0	95,136		46.8

Degree of thr damage

Slight damage. The damage has not effected the silvicultural quality of the stand.

Moderate damage. The damage has decreased the silvicultural quality of the stand by one class or made a low-yielding stand even less productive. The damage has not changed the development class (except possibly destroyed the upper storey over an established seedling storey).

Severe damage. The damage has decreased the silvicultural quality of the stand by more than one class or changed the development class into a temporarily unstocked stand or made a low-yielding stand significantly less productive.

Complete damage. The stand must be regenerated immediately.

The degree of the damage is a stand level variable describing the summed effect of all the damages observed. The base line for the estimation is the state of the stand before the damage. The effect of the damages on the growth and yield, mortality and quality of timber is the main criteria for the degree of the damage

Appendix

Table A.35 The areas of damaged forests by symptom description and by degree of damage on forest land in whole country

Symptom description	Degree of damage								Total		Proportion of forest land area
	Slight		Moderate		Severe		Complete				
	km²	%	km²	%	km²	%	km²	%	km²		%
Dead standing trees	4,725	10.5	4,851	12.4	1,803	17.9	170	23.4	11,549		5.7
Fallen or broken trees	3,604	8.0	1,837	4.7	635	6.3	55	7.6	6,131		3.0
Decayed standing trees	2,060	4.6	3,839	9.8	1,103	10.9	48	6.6	7,050		3.5
Stem or root damage	3,134	6.9	1,335	3.4	248	2.5	30	4.1	4,747		2.3
Resin flows	91	0.2	22	0.1	0	0.0	0	0.0	114		0.1
Dead or broken tops	1,915	4.2	1,671	4.3	648	6.4	86	11.8	4,319		2.1
Other top damage	9,736	21.6	9,997	25.5	1,873	18.5	154	21.2	21,760		10.7
Stem deformation	12,202	27.0	13,056	33.3	1,781	17.6	66	9.1	27,105		13.3
Branch damage	426	0.9	124	0.3	24	0.2	1	0.2	575		0.3
Abnormally pruned crown	2,270	5.0	547	1.4	38	0.4	0	0.0	2,855		1.4
Defoliation	4,085	9.0	1,687	4.3	174	1.7	3	0.4	5,949		2.9
Discolouration	904	2.0	190	0.5	18	0.2	0	0.0	1,112		0.5
Multiple symptoms	0	0.0	6	0.0	1,751	17.3	114	15.7	1,871		0.9
Total	45,152	100.0	39,161	100.0	10,096	100.0	728	100.0	95,136		46.8

Table A.36 The proportions of the defoliation target trees by defoliation and age classes

	Scots pine					Norway spruce				
	Age class, year					Age class, year				
	−39	40–79	80–119	120–	Total	−39	40–79	80–119	120–	Total
	Proportion of trees in age class, %					Proportion of trees in age class, %				
South Finland										
0–10%	89.6	66.8	44.1	30.6	61.7	96.6	61.1	27.7	19.2	50.2
11–25%	9.4	29.3	47.5	53.1	32.4	3.2	33.5	51.1	45.8	37.4
26–60%	1.0	3.8	8.0	15.8	5.7	0.3	5.1	20.4	32.5	11.9
over 60%	0.1	0.1	0.4	0.5	0.2	0.0	0.2	0.8	2.5	0.6
Age class total	100.0	100.0	100.0	100.0	100.0	100.0	100.0	100.0	100.0	100.0
North Finland										
0–10%	91.3	75.7	54.9	22.8	58.6	97.1	82.0	43.7	16.0	37.1
11–25%	7.9	20.7	36.9	49.0	30.6	2.9	14.6	42.5	36.4	32.8
26–60%	0.7	3.5	7.9	25.7	10.0	0.0	3.4	13.2	41.0	26.4
over 60%	0.1	0.1	0.3	2.5	0.8	0.0	0.0	0.6	6.6	3.8
Age class total	100.0	100.0	100.0	100.0	100.0	100.0	100.0	100.0	100.0	100.0
Whole country										
0–10%	90.2	71.4	49.2	25.1	60.1	96.6	64.1	31.0	16.9	46.6
11–25%	8.9	24.9	42.4	50.3	31.5	3.2	30.8	49.3	38.9	36.1
26–60%	0.9	3.7	8.0	22.7	7.9	0.3	4.9	18.9	38.8	15.9
over 60%	0.1	0.1	0.4	1.9	0.5	0.0	0.2	0.8	5.5	1.5
Age class total	100.0	100.0	100.0	100.0	100.0	100.0	100.0	100.0	100.0	100.0

Table A.37 The volume of dead wood on forest land and poorly productive forest land

Tree species group	Standing trees Mean volume m³/ha	s.e.	Total volume 1,000 m³	s.e.	Lying trees Mean volume m³/ha	s.e.	Total volume 1,000 m³	s.e.	Total Mean volume m³/ha	s.e.	Total volume 1,000 m³	s.e.
South Finland												
Pine	0.35	0.02	4,067	176	0.75	0.03	8,661	306	1.10	0.03	12,728	369
Norway spruce	0.22	0.02	2,514	179	0.52	0.02	6,010	239	0.73	0.03	8,523	314
Birch	0.12	0.01	1,408	104	0.26	0.01	3,001	124	0.38	0.02	4,409	177
Aspen	0.02	0.00	262	57	0.07	0.01	853	96	0.10	0.01	1,115	118
Other broadleaved	0.08	0.01	875	64	0.09	0.01	991	69	0.16	0.01	1,866	105
Unidentified coniferous	0.00	.	12	7	0.09	0.01	1,044	77	0.09	0.01	1,056	78
Unidentified broadleaved	0.00	0.00	32	16	0.01	0.00	165	27	0.02	0.00	197	32
Unidentified tree species	0.00	.	6	4	0.17	0.01	2,027	135	0.18	0.01	2,034	136
Total	0.79	0.03	9,175	297	1.96	0.04	22,753	499	2.75	0.05	31,928	612
North Finland												
Pine	1.16	0.08	13,223	888	4.11	0.15	46,914	1,745	5.27	0.19	60,137	2,126
Norway spruce	0.33	0.03	3,761	346	1.01	0.07	11,510	812	1.34	0.09	15,272	988
Birch	0.30	0.02	3,480	221	0.59	0.03	6,729	308	0.89	0.04	10,209	426
Aspen	0.02	0.00	238	50	0.06	0.01	719	114	0.08	0.01	957	133
Other broadleaved	0.03	0.00	287	43	0.03	0.00	343	44	0.06	0.01	630	70
Unidentified coniferous	0.00	0.00	26	13	0.32	0.02	3,677	279	0.32	0.02	3,703	280
Unidentified broadleaved	0.00	.	4	3	0.02	0.01	262	59	0.02	0.01	266	60
Unidentified tree species	0.00	.	12	4	0.30	0.03	3,456	311	0.30	0.03	3,468	311
Total	1.84	0.09	21,031	1034	6.45	0.17	73,610	1994	8.29	0.22	94,641	2,525

(continued)

Table A.37 (continued)

Tree species group	Standing trees				Lying trees				Total			
	Mean volume m³/ha	s.e.	Total volume 1,000 m³	s.e.	Mean volume m³/ha	s.e.	Total volume 1,000 m³	s.e.	Mean volume m³/ha	s.e.	Total volume 1,000 m³	s.e.
Whole country												
Pine	0.75	0.04	17,290	905	2.42	0.08	55,575	1,772	3.17	0.09	72,865	2,157
Norway spruce	0.27	0.02	6,275	390	0.76	0.04	17,520	847	1.03	0.05	23,795	1,036
Birch	0.21	0.01	4,887	244	0.42	0.01	9,730	332	0.63	0.02	14,618	461
Aspen	0.02	0.00	500	76	0.07	0.01	1,572	149	0.09	0.01	2,072	178
Other broadleaved	0.05	0.00	1,162	78	0.06	0.00	1,334	82	0.11	0.01	2,496	126
Unidentified coniferous	0.00	.	38	15	0.21	0.01	4,721	289	0.21	0.01	4,758	291
Unidentified broadleaved	0.00	.	36	17	0.02	0.00	427	65	0.02	0.00	463	67
Unidentified tree species	0.00	.	18	6	0.24	0.01	5,483	339	0.24	0.01	5,502	339
Total	1.31	0.05	30,207	1,076	4.19	0.09	96,363	2,055	5.50	0.11	126,569	2,598

Table A.38 The mean volume and proportions of dead wood on forest and poorly productive forest land

Tree species group	Standing trees −30 cm	Standing trees Over 30 cm	Lying trees −30 cm	Lying trees Over 30 cm	Mean volume m³/ha
	Proportion of the mean volume of the tree species group %				
South Finland					
Pine	28	4	54	14	1.10
Norway spruce	25	4	61	10	0.73
Birch	28	4	63	5	0.38
Aspen	17	7	56	21	0.10
Other broadleaved	46	1	52	1	0.16
Unidentified coniferous	0	1	82	17	0.09
Unidentified broadleaved	13	3	68	16	0.02
Unidentified tree species	0	0	84	16	0.18
Total	25	4	60	11	2.75
North Finland					
Pine	16	6	58	20	5.27
Norway spruce	20	4	55	21	1.34
Birch	32	2	63	3	0.89
Aspen	18	6	57	19	0.08
Other broadleaved	45	1	54	1	0.06
Unidentified coniferous	0	0	83	16	0.32
Unidentified broadleaved	2	0	89	9	0.02
Unidentified tree species	0	0	89	11	0.30
Total	17	5	60	17	8.29
Whole country					
Pine	18	6	58	19	3.17
Norway spruce	22	4	57	17	1.03
Birch	31	3	63	4	0.63
Aspen	17	7	56	20	0.09
Other broadleaved	46	1	52	1	0.11
Unidentified coniferous	0	0	83	16	0.21
Unidentified broadleaved	7	1	80	12	0.02
Unidentified tree species	0	0	87	13	0.24
Total	19	4	60	16	5.50

Table A.39 The total volume of dead wood by appearance classes on forest and poorly productive forest land

Appearance class	Pine	Norway spruce	Birch	Aspen	Other broadleaved	Unidentified coniferous	Unidentified broadleaved	Unidentified species	Total	Mean volume m³/ha
	Total volume, 1,000 m³									
South Finland										
Standing trees										
Whole tree	3,576	2,121	775	117	547	0	30	0	7,167	0.62
Snag	471	363	623	134	321	12	1	5	1,932	0.17
Broken tree	11	11	2		4	0	0	0	28	0.00
Cut stump or snag	9	19	7	10	3	0	0	1	49	0.00
Total	4,067	2,514	1,408	262	875	12	32	6	9,175	0.79
Lying trees										
Unknown, advanced decayed	1,180	200	274	17	11	480	41	1,320	3,523	0.30
Uprooted	3,351	2,655	614	157	196	265	17	244	7,500	0.65
Broken tree	2,137	1,522	1,192	358	352	165	35	277	6,037	0.52
Cut butt end or bolt	1,142	1,220	588	166	162	95	51	153	3,577	0.31
Logging waste	852	414	332	155	270	39	21	34	2,115	0.18
Total	8,661	6,010	3,001	853	991	1,044	165	2,027	22,753	1.96
Dead wood total	12,728	8,523	4,409	1,115	1,866	1,056	197	2,034	31,928	2.75
North Finland										
Standing trees										
Whole tree	11,464	3,447	2,201	169	173	0	0	0	17,454	1.53
Snag	1,747	311	1,261	69	114	23	4	12	3,541	0.31
Broken tree	6	2	0	0	0	0	0	0	7	0.00
Cut stump or snag	6	2	18	0	0	3	0	0	29	0.00
Total	13,223	3,761	3,480	238	287	26	4	12	21,031	1.84

Appendix

Lying trees									
Unknown, advanced decayed	5,045	625	701	77	12	1,365	64	2,096	9,986	0.87
Uprooted	21,176	3,586	961	183	57	860	17	489	27,329	2.39
Broken tree	10,856	6,566	4,240	363	182	1,152	175	723	24,257	2.13
Cut butt end or bolt	4,964	537	482	57	29	218	2	112	6,402	0.56
Logging waste	4,873	196	344	39	63	82	3	36	5,636	0.49
Total	46,914	11,510	6,729	719	343	3,677	262	3,456	73,610	6.45
Dead wood total	60,137	15,272	10,209	957	630	3,703	266	3,468	94,641	8.29
Whole country									
Standing trees									
Whole tree	15,041	5,569	2,975	287	720	0	30	0	24,621	1.07
Snag	2,218	674	1,885	204	435	34	6	17	5,473	0.24
Broken tree	17	12	2	0	4	0	0	0	35	0.00
Cut stump or snag	14	21	25	10	3	3	0	1	77	0.00
Total	17,290	6,275	4,887	500	1,162	38	36	18	30,207	1.31
Lying trees									
Unknown, advanced decayed	6,224	825	976	94	23	1,845	106	3,416	13,509	0.59
Uprooted	24,527	6,241	1,576	340	253	1,125	34	732	34,829	1.51
Broken tree	12,993	8,087	5,432	721	534	1,317	210	1,000	30,295	1.32
Cut butt end or bolt	6,106	1,757	1,071	223	191	313	53	265	9,979	0.43
Logging waste	5,724	609	676	194	333	121	24	70	7,751	0.34
Total	55,575	17,520	9,730	1,572	1,334	4,721	427	5,483	96,363	4.19
Dead wood total	72,865	23,795	14,618	2,072	2,496	4,758	463	5,502	126,569	5.50

Table A.40 The total volume of dead wood by tree species groups and by decay classes on forest land

	Decay class						
	1			2			3
	Standing trees	Lying trees	Total	Standing trees	Lying trees	Total	Standing trees
Tree species group	Total volume, 1,000 m³						
South Finland							
Pine	3,248	1,652	4,900	653	1,318	1,971	101
Norway spruce	2,309	1,987	4,296	157	1,114	1,271	40
Birch	370	589	959	416	460	876	376
Aspen	136	203	339	72	143	215	46
Other broadleaved	355	282	637	258	207	465	171
Unidentified coniferous	7	19	26	5	75	80	0
Unidentified broadleaved	20	8	28	0	16	16	11
Unidentified tree species	1	55	55	1	109	110	1
Total	6,446	4,794	11,240	1,562	3,443	5,005	745
Mean volume of the decay class, m³/ha	0.56	0.41	0.97	0.13	0.30	0.43	0.06
North Finland							
Pine	9,570	3,432	13,002	2,896	6,549	9,445	681
Norway spruce	3,478	1,706	5,183	173	1,836	2,009	66
Birch	774	584	1,358	1,073	852	1,925	1,058
Aspen	198	155	353	24	106	130	16
Other broadleaved	68	66	133	111	102	213	75
Unidentified coniferous	9	56	65	11	361	372	6
Unidentified broadleaved	0	18	18	4	12	17	0
Unidentified tree species	9	54	63	1	218	219	0
Total	14,106	6,070	20,176	4,294	10,036	14,330	1,902
Mean volume of the decay class, m³/ha	1.24	0.53	1.77	0.38	0.88	1.26	0.17
Whole country							
Pine	12,818	5,084	17,903	3,549	7,867	11,416	782
Norway spruce	5,787	3,692	9,479	329	2,950	3,279	106
Birch	1,144	1,172	2,317	1,489	1,312	2,801	1,433
Aspen	334	358	692	96	250	345	62
Other broadleaved	423	348	771	369	309	679	246
Unidentified coniferous	16	75	91	16	436	452	6
Unidentified broadleaved	20	26	47	4	29	33	11
Unidentified tree species	10	108	118	2	327	329	1
Total	20,552	10,864	31,416	5,856	13,479	19,335	2,647
Mean volume of the decay class, m³/ha	0.89	0.47	1.37	0.25	0.59	0.84	0.12

Standing trees, decay classes:

1 Hard. A knife penetrates only a few millimetres into the wood. Bark (almost) intact, branches remain. Also includes hard barkless trees, the wood of which has not started to decay

2 Fairly hard. A knife penetrates 1–2 cm into the wood. Branches have started to fall off, bark of coniferous trees has started to peel off, broadleaved species often host multiple fruiting bodies of polypores on the upper part of the tree

3 Fairly soft. A knife penetrates 3–5 cm into the wood. With coniferous species, bark only left at the base of the stem, with broadleaved species the bark remains but the trunk has started to decay, branches have mainly fallen off, only larger branches remain but are not intact, part of the crown has often collapsed

4 Soft. A knife easily penetrates deep into the wood. The trunk remains standing only because of support from the bark, branches have totally fallen off of broadleaved species, the tree is usually broken - only the stem snag is standing

Appendix

and poorly productive forest land

Lying trees	Total	4 Standing trees	Lying trees	Total	5 Lying trees	Total Standing trees	Lying trees	Total
1,801	1,901	65	2,276	2,341	1,615	4,067	8,661	12,728
1,128	1,168	8	1,212	1,220	570	2,514	6,010	8,523
514	889	246	786	1,032	652	1,408	3,001	4,409
135	181	8	254	262	118	262	853	1,115
202	373	91	199	290	100	875	991	1,866
205	205	0	308	308	438	12	1,044	1,056
26	38	0	40	40	75	32	165	197
249	249	4	563	567	1,052	6	2,027	2,034
4,260	5,004	422	5,637	6,059	4,619	9,175	22,753	31,928
0.37	0.43	0.04	0.49	0.52	0.40	0.79	1.96	2.75
10,898	11,580	76	15,724	15,800	10,311	13,223	46,914	60,137
2,897	2,963	45	3,620	3,665	1,452	3,761	11,510	15,272
843	1,901	574	2,278	2,852	2,173	3,480	6,729	10,209
211	227	0	159	159	87	238	719	957
121	196	33	38	71	17	287	343	630
549	555	0	1,152	1,152	1,559	26	3,677	3,703
24	24	0	98	98	109	4	262	266
438	438	2	996	997	1,751	12	3,456	3,468
15,982	17,884	729	24,065	24,794	17,458	21,031	73,610	94,641
1.40	1.57	0.06	2.11	2.17	1.53	1.84	6.45	8.29
12,699	13,481	141	18,000	18,140	11,925	17,290	55,575	72,865
4,025	4,131	53	4,831	4,885	2,022	6,275	17,520	23,795
1,357	2,790	820	3,064	3,884	2,825	4,887	9,730	14,618
347	409	8	413	421	205	500	1,572	2,072
324	569	124	237	361	117	1,162	1,334	2,496
753	759	0	1,460	1,460	1,996	38	4,721	4,758
50	61	0	138	138	184	36	427	463
687	688	6	1,559	1,564	2,803	18	5,483	5,502
20,241	22,888	1,152	29,702	30,853	22,077	30,207	96,363	126,569
0.88	0.99	0.05	1.29	1.34	0.96	1.31	4.19	5.50

Lying trees, decay classes:
1 Hard. A knife penetrates only a few millimetres into the wood. A recently fallen trunk with bark remaining, possible epiphytes are associated with standing trees. Also includes hard trunks that have fallen down long after dying but the wood has not started to decay
2 Fairly hard. A knife penetrates 1–2 cm into the wood. The bark often remains, only a few epiphytes that are mostly associated with standing trees
3 Fairly soft. A knife penetrates 3–5 cm into the wood. The bark is often torn and peeled, locally abundant epiphytes but small growths. The category often includes e.g. pines, when sapwood is advanced decayed and only heartwood is hard
4 Soft. A knife easily penetrates deep into the wood. Often barkless and covered by epiphytes. Large growths of mosses and lichens
5 Very soft. Disintegrates on being touched. Usually entirely covered by epiphytes, most epiphytes (mosses, lichens and shrubs) associated with forest ground, only a sligth bulge distinguishes trunk from the surrounding ground

Table A.41 The areas of key habitats by ecological value and by protection status on forest land, poorly productive forest land and unproductive land

Key habitat	Protected area or planned to be protected area				Not protected				Total			
	Ecological value				Ecological value				Ecological value			
	Not valuable	Valuable	Forest Act	Total	Not valuable	Valuable	Forest Act	Total	Not valuable	Valuable	Forest Act	Total
	km²											
Spring	0	0	2	2	7	27	16	50	7	27	17	52
Groundwater surfaces without a visible spring	0	1	5	6	15	4	4	23	15	4	9	28
Brook-side forest	2	27	73	102	76	109	122	307	78	136	195	409
Stand surrounding a small (<1ha) pond	0	55	36	91	27	102	47	175	27	157	83	266
Fen or bog surrounding a small pond	0	29	47	76	5	42	82	130	5	71	130	206
Other small wetland area	2	7	11	20	28	26	41	95	29	34	52	115
Eutrophic paludified hardwood-spruce forest	9	62	18	89	410	150	78	639	419	212	97	728
Eutrophic birch or hardwood-spruce fen	0	131	27	158	107	229	34	370	107	360	61	528
Eutrophic pine fen	7	150	16	173	313	360	25	697	320	510	40	870
Mesotrophic hardwood-spruce mire	69	265	157	491	2,831	720	392	3,943	2,900	985	548	4,434
Oligotrophic hardwood-spruce and pine fens	58	1,017	55	1,130	1,053	2,356	56	3,466	1,111	3,374	110	4,595
Oligotrophic spruce mires and fens	26	229	16	271	173	297	14	484	199	526	30	756
Ombrotrophic pine bogs	148	1,749	157	2,053	4,668	4,020	132	8,820	4,816	5,769	289	10,873
Sphagnum fuscum-dominated bogs	30	288	76	394	644	886	133	1,663	674	1,174	210	2,058
Eutrophic and mesotrophic fens	0	329	0	329	30	357	13	400	30	686	13	729
Open fens and bogs	63	4,255	483	4,801	985	7,334	456	8,775	1,048	11,589	939	13,576
Alluvial meadow	14	78	32	124	61	175	128	364	75	253	160	488
Dry mesotrophic herb-rich forests	2	0	8	10	49	26	9	85	51	26	17	95

Dry eutrophic herb-rich forests	0	3	0	3	8	2	1	11	8	5	1	14
Mesic mesotrophic herb-rich forests	33	37	41	112	1,048	194	47	1,290	1,082	232	88	1,401
Mesic eutrophic herb-rich forests	19	20	12	51	500	169	69	737	518	189	80	788
Moist mesotrophic herb-rich forestst	8	1	7	16	196	77	28	301	204	78	35	317
Moist eutrophic herb-rich forests	8	20	36	64	369	173	52	594	377	193	88	659
Naturally regenerated rare hardwood forest	0	3	1	4	0	7	4	11	0	9	6	15
Islet (<1 ha) of mineral soil forest on undrained peatland	1	18	43	62	53	61	36	150	54	79	79	212
Gorge	0	0	1	1	0	0	0	0	0	0	1	1
Ravine	0	2	9	11	0	1	0	1	0	3	9	12
Precipice (>10 m high)	2	6	21	29	46	57	38	141	48	63	59	170
Open rock (soil layer at most patchy, only few trees)	14	52	97	163	361	572	692	1,625	375	624	789	1,788
Small formations of rock	0	0	2	2	0	9	3	12	0	9	5	13
Stonefields, boulder fields	12	574	157	743	105	411	145	661	117	985	302	1,404
Sand fields	0	0	19	19	6	0	4	10	6	1	22	28
Other rare biotope	0	0	0	0	0	8	9	17	0	8	9	17
Total	526	9,409	1,665	11,600	14,174	18,963	2,910	36,047	14,700	28,372	4,575	47,647
Proportion of forest land, poorly productive forest land and unproductive land, %	0.2	3.6	0.6	4.4	5.4	7.2	1.1	13.8	5.6	10.8	1.7	18.2

Table A.42 The areas of key habitats by naturalness class on forest land, poorly productive forest land and unproductive land

Key habitat	Naturalness Natural km²	Signs of human activity Present, not altered	Present, altered	Present, strongly altered	Total
Spring	22	12	14	5	52
Groundwater surfaces without a visible spring	6	7	4	11	28
Brook-side forest	172	104	76	57	409
Stand surrounding a small (<1 ha) pond	158	59	33	16	266
Fen or bog surrounding a small pond	161	27	16	3	206
Other small wetland area	51	33	16	16	115
Eutrophic paludified hardwood-spruce forest	112	79	128	409	728
Eutrophic birch or hardwood-spruce fen	274	125	48	81	528
Eutrophic pine fen	364	164	99	243	870
Mesotrophic hardwood-spruce mire	747	541	682	2,464	4,434
Oligotrophic hardwood-spruce and pine fens	2,383	1,029	641	542	4,595
Oligotrophic spruce mires and fens	327	243	99	87	756
Ombrotrophic pine bogs	3,480	2,416	2,075	2,903	10,873
Sphagnum fuscum -dominated bogs	945	344	279	490	2,058
Eutrophic and mesotrophic fens	602	87	16	24	729
Open fens and bogs	10,575	1,880	589	532	13,576
Alluvial meadow	242	139	66	42	488
Dry mesotrophic herb-rich forests	4	21	43	27	95
Dry eutrophic herb-rich forests	0	1	9	5	14
Mesic mesotrophic herb-rich forests	82	97	510	713	1,401
Mesic eutrophic herb-rich forests	35	93	260	400	788
Moist mesotrophic herb-rich forests	22	42	148	106	317
Moist eutrophic herb-rich forests	66	107	197	289	659
Naturally regenerated rare hardwood forests	4	7	5	0	15
Islet (<1 ha) of mineral soil forest on undrained peatland	68	61	61	22	212
Gorge	1	0	0	0	1
Ravine	11	1	0	0	12
Precipice (>10 m high)	43	55	55	17	170
Open rock (soil layer at most patchy, only few trees)	778	591	313	107	1,788
Small formations of rock	5	6	2	0	13
Stonefields, boulder fields	1,103	171	95	35	1,404
Sand fields	16	5	5	1	28
Other rare biotope	9	3	3	1	17
Total	22,868	8,544	6,586	9,649	47,647
Proportion of forest land, poorly productive forest land and unproductive land, %	8.7	3.3	2.5	3.7	18.2

Table A.43 Key habitat management on forest land, poorly productive forest land and unproductive land

Key habitat	Accomplished measures							Total
	0	1	2	3	4	5	6	
	km²							
Spring	16	1	2	1	0	0	32	52
Groundwater surfaces without a visible spring	18	0	0	1	0	0	8	28
Brook-side forest	130	16	23	11	12	1	216	409
Stand surrounding a small (<1ha) pond	41	6	3	2	3	0	211	266
Fen or bog surrounding a small pond	15	2	1	7	2	0	180	206
Oher small wetland area	24	22	0	7	2	0	60	115
Eutrophic paludified hardwood-spruce forest	490	18	2	13	0	0	205	728
Eutrophic birch or hardwood-spruce fen	121	0	0	10	1	0	396	528
Eutrophic pine fens	318	2	0	12	0	0	538	870
Mesotrophic hardwood-spruce mire	2,922	75	6	58	5	2	1,366	4,434
Oligotrophic hardwood-spruce and pine fens	821	45	1	66	10	0	3,653	4,595
Oligotrophic spruce mires and fens	107	0	0	9	0	0	640	756
Ombrotrophic pine bogs	3,858	96	15	151	18	9	6,727	10,873
Sphagnum fuscum -dominated bogs	621	16	3	68	5	5	1,340	2,058
Eutrophic and mesotrophic fens	33	3	0	2	0	0	690	729
Open fens and bogs	1,072	65	8	278	24	4	12,124	13,576
Alluvial meadow	139	1	3	19	4	0	323	488
Dry mesotrophic herb-rich forests	70	2	0	0	0	3	20	95
Dry eutrophic herb-rich forests	11	1	0	0	0	0	2	14
Mesic mesotrophic herb-rich forests	1,177	34	9	8	1	3	169	1,401
Mesic eutrophic herb-rich forests	620	31	4	9	0	3	121	788
Moist mesotrophic herb-rich forests	249	11	0	5	0	0	52	317
Moist eutrophic herb-rich forests	457	24	5	11	2	0	158	659
Naturally regenerated rare hardwood forests	9	3	0	2	0	0	1	15
Islet (<1 ha) of mineral soil forest on undrained peatland	35	3	0	2	1	0	170	212
Gorge	0	0	0	0	0	0	1	1
Ravine	0	1	0	0	0	0	11	12
Precipice (>10 m high)	82	13	4	9	2	0	59	170

(continued)

Table A.43 (continued)

Key habitat	Accomplished measures							Total
	0	1	2	3	4	5	6	
	km²							
Open rock (soil layer at most patchy, only few trees)	484	31	3	117	2	0	1,151	1,788
Small formations of rock	4	3	0	2	0	0	5	13
Stonefields, boulder fields	118	5	0	38	0	0	1,243	1,404
Sand fields	6	1	0	0	0	0	21	28
Other rare biotope	5	1	0	6	0	0	5	17
Total	14,075	533	93	923	95	30	31,898	47,647
Proportion of forest land, poorly productive forest land and unproductive land, %	5.4	0.2	0	0.4	0	0	12.2	18.2

Measures accomplished on key habitats: *0* Nature of the key habitat has not been taken into consideration during measures, *1* Careful operations inside key habitat, *2* Careful operations inside key habitat and the recommended buffer area, *3* Key habitat left untouched during management measures, *4* Key habitat and buffer left untouched, *5* Key habitat managed for enhancing habitat value (e.g., removal of spruce in herb-rich forests and rare hardwood stands), *6* No measures on the key habitat or in the surrounding stands during past 30 years

Table A.44 The occurrence of keystone tree species on forest land and on poorly productive forest land

Keystone tree species	Diameter threshold at breast height	stems/ha	1,000 stems	Proportion of total stem number %
South Finland				
European aspen	≥30 cm	0.56	6,489	10.1
Grey alder	≥20 cm	0.30	3,522	5.5
Black alder	≥10 cm	1.89	21,888	34.2
European mountain ash	≥10 cm	0.81	9,446	14.8
Goat willow	≥10 cm	1.55	18,024	28.2
Fluttering elm	≥5 cm	0.00	41	0.1
Wych elm	≥5 cm	0.00	17	0.0
Small-leaved lime	≥5 cm	0.11	1,294	2.0
European ash	≥5 cm	0.08	882	1.4
Pedunculate oak	≥5 cm	0.10	1,135	1.8
Hazel	≥5 cm	0.00	25	0.0
Norway maple	≥5 cm	0.11	1,221	1.9
Total		5.52	63,983	100.0
North Finland				
European aspen	≥30 cm	0.15	1,715	9.9
Grey alder	≥20 cm	0.03	346	2.0
Black alder	≥10 cm	0.03	379	2.2
European mountain ash	≥10 cm	0.06	741	4.3
Goat willow	≥10 cm	1.24	14,147	81.6
Total		1.52	17,327	100.0

(continued)

Appendix

Table A.44 (continued)

Keystone tree species	Diameter threshold at breast height	stems/ha	1,000 stems	Proportion of total stem number %
Whole country				
European aspen	≥30 cm	0.36	8,204	10.1
Grey alder	≥20 cm	0.17	3,868	4.8
Black alder	≥10 cm	0.97	22,267	27.4
European mountain ash	≥10 cm	0.44	10,187	12.5
Goat willow	≥10 cm	1.40	32,171	39.6
Fluttering elm	≥5 cm	0.00	41	0.0
Wych elm	≥5 cm	0.00	17	0.0
Small-leaved lime	≥5 cm	0.06	1,294	1.6
European ash	≥5 cm	0.04	882	1.1
Pedunculate oak	≥5 cm	0.05	1,135	1.4
Hazel	≥5 cm	0.00	25	0.0
Norway maple	≥5 cm	0.05	1,221	1.5
Total		3.53	81,310	100.0

Index

Note: Depending on the type of the key word, **bold** page number can refer to the definition of the term, to the explanation of the assessment of a variable in NFI9, or to the table of variable values in which the word occurs.

A
Abies *See* Fir
Abiotic damage, 153
Administrative information, **28–29**
Afforestation, 99, 106
Age, 46, 54, 64, 69, 122, 123, 130–131, 146
    distribution, 121–124
    distribution on protected areas, 149, 150
    mean, by development class, 207, 208
    tree, 51, **54**, 60, 63
    tree storey, 37, 40, **41** (*see also* Development class)
    uneven, 37, **38** (*see also* Tree storey)
Age core *See* Core
Agricultural land*See* Land use
Air pollution, 43, 55
Åland, 7, 20, 137
Alder *See* Black alder; Grey alder
Alectoria, 55
Alnus glutinosa See Black alder
Alnus incana See Grey alder
Angle count plot *See* Sample plot
Archaeological site, 29
Area
    estimation of, 2, 69, **70–71**, 88
    sampling error of, 87, **87**, 89
Ash *See* European ash; European mountain ash
Aspen, **52**, 55, **57**, 116, 132
    dead, 168, 169, 170
    diameter distribution, 223

    as keystone species, 260
    volume by diameter class, 224
    volume function, 75, **75–77**
Assortment *See* Timber assortment

B
Bark *See* Tree
Basal area, 40, **41**, 45, 69, 71, **72, 74**, 81
    by development class, 206
    tree, **71**, 74, 75, 80
Basal area factor, 23, 24, **25**, 41, 72, 87
    See also Sample plot, angle count
Bay willow, 50, **52**
Bedrock, 34, **34**, 36, 101, 222 *See also* Soil
Bias, 5, 70, 73, 76
Binoculars, 53, **59**, 60
Biodiversity, 11, 35, 36, 38, 46, 49, 55, 56, 148, 162–174, 182
Bio-energy, 12
Biomass, 11
Birch, **57** *See also* Downy birch; Silver birch
    dead, 168, 169, 170
    as dominant species, 117, 131, 198
    growing stock of, 132
    increment of, 80, 82, 84, 145
    mean volume, 126, 128
    mean volume, by development class, 206
    volume function, 75, **75–77**
Bird cherry, **52**, 222
Bi-stand, **26** *See also* Sample plot, divided

Bitterlich plot *See* Sample plot, angle count
Black alder, 33, **52**, 56, **57**, 116, 171
    as keystone species, 259
    volume function, **75–77**
Bog *See* Peatland
Boulder field, 34, **48**, 101
Branch *See* Tree
Breast height, **50**, 51
Breast height diameter *See* Diameter
Breeding forest, 29
Broad-leaved species *See also* Birch;
    Tree species
    drain, 84
    increment of, 84
    mean volume, 129
Broad-leaved species (other than birch)
    as dominant, 116
    growing stock of, 89
    increment of, 80, 82
    mean volume, 126, 129, 216
    mean volume, by development class, 216
Brook, **48**, 222
Bryoria, 55
Bucking *See* Tree, bucking of
Budget, 65
Buffer zone, 37, **49**
Built-up land *See* Land use
Bush, 30, 49, **50**

C
Calculation region, **69**, 70, 73, 80, 83, 84, 85
    aggregates of, 70, 72, 74, 85, 86
Calliper, 59, **59**
Canopy cover, 30, 31
Centre point stand, **26**, 27, 40 *See also* Sample
    plot, centre of
Certification *See* Forest certification
Cherry *See* Bird cherry
Clear cut *See* Cutting; Regeneration,
    temporarily unstocked area
Climatic variation, 32, 139
    See also Temperature
Cluster, **17**, 20, 21, 22, 23, 64, 65, 86, 183
    See also Sampling design
    locations of, **18–19**, 180
Compass, 27, **59**
Coniferous species *See* Tree species
Control *See* Measurement
Core, 41, 54, 60, 64, 138, 183
Cost *See* Measurement, costs of
COST Action E43, 49, 162
Crew leader *See* Field crew
Crown *See* Tree
Crown cover *See* Canopy cover

Cut point, **54, 55**
Cutting, 2, 45, 159 *See also* Drain; Forest
    management; Silvicultural measure;
    Wood production
    accomplished, **45**, 158
    possibilities, 31
    proposed, 39, **45**, 159
    regeneration, 37, 40
    regime, 38, 47, 158, 172
    scenario, 46
    thinning, 38, 40, 158
    timing of, 159, 237

D
Damage, 36, 38, **42–44**, 45, 150–154
    age of, **43**, 150
    causal agent of, 43, **43**, 152
    degree of, 44, **44**, 150
    symptom of, **42**, 151, 243
    tree, 51, **54**
Dead wood, 27, 36, 165–172
    appearance of, **57**, 251
    class of, **53**
    coarse, 165, 169
    decay class, 58, **58**, 169–170, 252
    by development class, 170
    lying, **50, 56**, 57, **57**, 85, 166, 168, 170
    mean volume, 168, 169, 170
    measurements on, **56–58**, 182
    standing, **50, 56**, 57, **57**, 85, 168, 169
    usable, **50**, 56, 171, 172, 173
    volume, 69, **85**, 166, 168, 179, 247–249
Deciduous species *See* Tree species,
    broad-leaved
Defoliation, 42, 51, **54**, 154, 246
Deformation *See* Stem deformation
Design *See* Sampling design
Desmococcus olivaeus, 55
Development class, 37, 38, **40**, 44, 124–126,
    132, 170, 174, 206–211
    mature, 32, 38, **40**, 125, 171
    regeneration stand (*see* Regeneration,
    temporarily unstocked area)
    seedling stands, 11, 37, 38, 40, **40**, 41, 124
    seed tree stand, **40**, 45, 124
    shelter tree stand, **40**, 45, 124
    thinning stands, 38, **40**, 124
    (*see also* Cutting, thinning)
Diameter, 40
    class, 80, 82, 85, 134
    dead wood, 56, 85
    distribution, 202
    mean, **41**, 69, 71, **71**, 72, 74
    mean, by development class, 206–210

Index 263

tree, 25, 26, 51, **51**, 55, 59, 60, 71, 75, 78, 79, 81
upper, 51, **53**, 59, 75, 78
Diameter increment, 12, 51, **54**, 60, 64, 80, 81, 138, 183
Die-back, 42
Digital elevation model *See* Elevation
Discolouration, 42
Ditch *See* Drainage
Divided plot *See* Sample plot, divided
Dominant species *See* Tree species, dominance
Dominant storey *See* Tree storey
Double sampling *See* Sampling design
Downy birch, **52, 57** *See also* Birch; Broad-leaved species
diameter distribution, 223
growing stock of, 89, 213–221, 223, 224
mean volume, 213–218
stem number, 222
Drain, 133, 140, 145
increment of, 80, **83–84**
volume of, **84**, 179–180
Drainage, 10, 11, 32, **35**, 36, 46, **46**, 80, 82, 108, 112, 160–162, 179
See also Transformed site; Transforming site
complementary or ditch cleaning, 162
erroneous, **46**
first-time, 161
prospective needs for, **46**

E
Earth drill, **59**
Edge effect, 73
Effective temperature sum *See* Temperature
Elevation, **31**, 101 *See also* Topography
digital elevation model, 31
Elm *See* Fluttering elm; Wych elm
English yew, **52**
Epiphyte *See* Lichen
Establishment, 45 *See also* Stand, establishment type; Tree, origin type of
European ash, **52**, 56, 259
European aspen *See* Aspen
European mountain ash, **52**, 56, **57**
as keystone species, 259, 260

F
FAO class *See* Land class (FAO)
Fell, 32
Felling *See* Cutting
Fen *See* Peatland
Fertility *See* Site fertility

Field computer, 28, **59**, 60
Field crew, 61, 63, 64, 65
training of, **63–64**, 65
Field plot *See* Sample plot
Finnish Broadcasting company, 27
Finnish Environment Institute, 29
Finnish Meteorological Institute, 24
Fir, **52**, 221
Fire, damage caused by, 43, 153
Fixed radius plot *See* Sample plot
Fluttering elm, **52**, 56
as keystone species, 259, 260
Food and Agriculture Organization (FAO), 9
Forest *See also* Breeding forest; Forest land; Land use; Research forest
FAO definition (*see* Land class (FAO))
old-growth, 56, 123, 148
small, 31, 46, 47, **48**
Forest Act, 9, 47, 49, 163, 180, 182
Forest balance, 140–143, 179
Forest certification, 47
Forested area, 32, 188, 189
Forest industry, 2, 8
strategic planning, 1
Forest land, 25, **30**, 33, 35, 50, 69, 71, 80, 82, 85, 99, 129
area of, 70, 71, 73, 95–98
sampling error in area, 89, 181
Forest management, 29, 36, 39, 45, 47, 49, 148, 154, 171 *See also* Cutting; Silvicultural measure; Wood production
large-area planning, 9
Forest Resources Assessment (FRA), 9
Forest road, 30, 31 *See also* Land use
Forestry centre, 9, **18–19**, 56, 57, 65, 69, 137, 180
Forestry land *See also* Land use
area of, 93, 95, 96
Forestry measure *See* Silvicultural measure
Fork tree, **50**, 53, 55
Form factor *See* Tree
Form height *See* Tree
Fraxinus excelsior *See* European ash
Fuel wood, 84

G
Geodetic Institute, 28
Glacial till, **34**, 101, 196 *See also* Soil
Global positioning system, 27, **59**
accuracy of, 28, 183
Goat willow, 50, **52**, 56, **57**
as keystone species, 259, 260

Gorge, **48**
Grain size *See* Soil, grain size
Greenhouse gas reporting, 10
Grey alder, **52**, 55, **57**, 132
    as keystone species, 259, 260
    volume function, 75, **75–77**
Ground level, **50**
Growing season, 2–4, 12, 80, 83
Growing stock *See* Volume; Volume
    increment
Growth, 35, 44, 53, 80 *See also* Diameter
    increment; Height increment;
    Volume increment
    index, 139, 140
    variation of, 54, 139, 140
Growth space, **54**

H
Habitat *See* Dead wood; Key habitat;
    Protected area
Harvest, 83, 85 *See also* Drain; Silvicultural
    measure; Wood production
Height, 30, 40
    dead wood, 56, 85
    mean, 37, **41**
    tree, 51, **53**, 59, 62–64, 75, 81, 182
Height increment, 51, **53**, 59, 62–64, 81, 138,
    139, 182, 880
Herb-rich forest, 29, **32**, 47, 48, 102,
    148, 164 *see also* site fertility
Heterobasidion annosum, 154
HKLN, 9
Horwitz-Thompson estimator, 73
Hummock, 33
Humus, **35**, 104 *See also* Soil, organic layer
Hydrology, 33, 35
Hypogymnia, 55
Hypsometer, 59, **59**, 60, 182

I
Inclusion probability *See* Sampling design
Increment *See* Diameter increment;
    Height increment; Volume
    increment
Increment borer, **59**
Increment core *See* Core
Increment measurement period, 80, **80**,
    81, 83
Insect, damage caused by, 43, 153
International GNSS Service, 28
Inundation, **33**
Isostatic recovery, 93

J
Juniper, 50, **52**
Juniperus communis See Juniper

K
Key biotope *See* Key habitat
Key habitat, 11, 26, **46–49**, 163–165, 180, 182
    accomplished measures on, 49, **49**
    altered, **49**, 164
    area of, 47, 69, **70–71**
    class of, **48**
    ecological value of, 49, **49**, 163
    group, 163
    naturalness of, **48, 49**, 163, 164
    restoration, 163
Keystone species, 27, **55–56**, 171–172, 259–260
k nearest neighbour method (k-NN), 78
Knife, 58, **59**
Kriging, 24, 89
Kyoto Protocol, 1, 10–11

L
Lake *See* Shore; Water
Land area, 70, 86, 93, 94
Land class (FAO), **30–31**, 98
    See also Land use
Landsat 5 TM *See* Satellite image
Land use, 20, **30–31**, 33, 69, 82
    See also Forest land; Land class
    (FAO); Poorly productive forest
    land; Unproductive land; Water
    change in, 31, 52, 98–99
    forestry, 30, **30**, 69, 89
Land Use, Land-Use Change and Forestry
    (LULUCF), 9, 10
Land-use plan, 29
Larch, **52**
    volume function, 75, **75,** 76
Larix See Larch
Length
    broken part of tree, 53
    dead wood, 56
    quality part (timber assortment), 54
    tree (*see* Height)
Lichen, 11, **55**, 58
Lime *See* Small-leaved lime
Limestone, 47
Line survey, 5, 87
Location *See* Sample plot, locating
Lodgepole pine, **52**
Low-yielding *See* Silvicultural quality,
    low-yielding

Index

M
Management *See* Forest management
Map data, 8, 20, 23, 24, 27, 29, 69, **89–90**, 180
Maple *See* Norway maple
Maturity *See* Development class, mature; Regeneration
Maximum radius *See* Sample plot
Meadow species, 33, 48
Measurement
    accuracy of, 62–63, **63–64**
    costs of, 24, 25, **64–65**, 182
    instruments, **58–60**, 63, 65, 182–183
Measurement pole, 53
Measurements on, 42, 75
Measuring rod, 60
Measuring tape, 27, **59**, 60
MELA, 10
Metsähallitus, 28, 29
Microscope, 41, 60, 64
Military area, 29
Mineral soil, 11, **31, 32**, 80, 82, 161
    See also Site class; Soil, mineral layer
Ministerial Conference on the Protection of Forests, 9
Mire *See* Peatland
Mire conservation, 148 *See also* Nature conservation; Peatland; Protected area; Wood production, restrictions on
Mixed forest, 119–120
Moder, **35**
Moisture, 33, 36
Molinia caerulea, 33
Moose, damage caused by, 43, 153
Mortality, 83 *See also* Dead wood; Drain
Mull soil, 34, **35**
Multi-source national forest inventory, 2, 8, 12, 20, 24, 27, 180
Municipality, 5, 8, **12**, 28, 180

N
National Forestry Program, 10
National Land Survey, 23, 24, 28, 31
National park, 29, 148, 149
Natural losses, **84**, 85, 133
    See also Dead wood; Drain
Naturalness, 35, 46, 164 *See also* Key habitat
Nature conservation, 29, 149 *See also* Wood production, restrictions on
Nature Conservation Act, 29, 47
Nature reserve, 29, 148 *See also* Wood production, restrictions on

Nearest neighbour methods *See* k nearest neighbour method
Needle loss, 42
Nitrogen deposition, 104
Norway maple, **52**, 56
    as keystone species, 259, 260
Norway spruce *See* Spruce
Number of trees *See* Stem number
Nurse crop, **39**
Nutrient *See* Site fertility

O
Oak *See* Pedunculate oak
Old-growth forest *See* Forest
Open bog *See* Peatland, treeless
Optimal allocation, 25
Organic layer *See* Soil, organic layer
Other broad-leaved species *See* Broad-leaved species (other than birch)
Over-aged stand *See* Silvicultural quality, reasons for decrease
Ownership, 28, **28**, 69, 99, 132, 148, 180

P
Paludification, 36, 48
Panel system, 183
Park, 30, 50
Parmelia, 55
Peat, **35**
Peat bore, **59**
Peatland, 11, **31**, 36, 80, 82, 104, 160, 163 *See also* Drainage; Site class
    area of, 104–106
    growing stock, 130, 132
    pine dominated, **32**, 106, 109, 110
    protected, 149
    restoration of, 46
    spruce dominated, **32**, 106, 109
    treeless, **32**, 48, 106, 110, 112, 149
    volume increment, 147–148
Peatland forest *See* Transformed site
Peatland moss, 33
Peat layer, 33, 106 *See also* Soil, organic layer
    thickness of, 111–112
Pedunculate oak, **52**, 56
    as keystone species, 259, 260
Permanent plot, 8, 11, 20, **20**, 23, 31, 34, 50, 51, **52**, 55, **56**, 63, 179
Petula pendula See Silver birch
Petula pubescens See Downy birch

Picea abies See Spruce
Pine, **52, 57** *See also* Lodgepole pine;
        Swiss stone pine
  damage, 154
  dead, 168, 169, 170
  diameter distribution, 223
  as dominant species, 113–118, 131
        (*see also* Peatland,
        pine dominated)
  drain, 84
  growing stock of, 89, 132, 213–221
  growth of, 140
  increment of, 80, 82, 84, 145
  mean volume, 126, 127, 129
  mean volume, by development class,
        206–211
  stem number, 222
  volume by diameter class, 224
  volume function, 75, **75–77**
Pine mire *See* Peatland,
        pine dominated
Pinus cembra See Swiss stone pine
Pinus contorta See Lodgepole pine
Pinus sylvestris See Pine
Pleurozium schreberi, 36
Plot *See* Sample plot
Pole size, 40, **41**
Pond, **48**, 164
Poorly productive forest land, 25, **30**, 33, 35,
        36, 50, 69, 71, 80, 82, 96
        See also Land use
  area of, 97, 98
  sampling error in area, 89
Poplar, **52**
Population Register, 28
Populus See Poplar
Populus tremula See Aspen
Potassium, deficiency of, 44
Power line *See* Land use
Precipice, **48**, 256–258
Precision *See* Sampling error
Principal site class *See* Site class
Productivity *See* Land use; Silvicultural
        quality; Site fertility; Wood
        production
  Protected area, 47, 126, 148–150, 163
        See also Key habitat; Nature
        conservation; Wood production,
        restrictions on
  by age class, 149, 150
Prunus padus See Bird cherry
Pseudevernia, 55
Pulpwood *See* Timber assortment
Pyrola, 33, 36

Q
Quality *See* Silvicultural quality;
        Technical quality
Quality assurance, **63–64**
Quality part, **54**
Quercus robur See Pedunculate oak

R
Ratio estimator, 70, 73, 85
  sampling error, **86–88**
Ravine, **48**
Recreation, 29, 100
Reforestation, 99, 106
Regeneration, 36, 37, 38, 39, 40, 44, 46
        See also Forest Management;
        Silvicultural measure; Stand,
        establishment type; Tree, origin
        type of; Wood production
  artificial, 109, 157, 159
  natural, 157
  success in, 38, 156–157
  temporarily unstocked area, 30, 38, 40, 44,
        45, 118
Register number, 28
Reindeer, 11, 43
Relascope, 26, **59** *See also* Sample plot,
        angle count
Removal *See* Drain
Research forest, 29, 189
Resin flow, 42
Restrictions *See* Wood production,
        restrictions on
Restroration *See* Peatland, restoration of
Retention tree, 36–37, **37**, 38, **39**, 40, 172, 174
River *See* Water
Road, 23, 24, 27 *See also* Forest road;
        Land use
Rock, 47, 48, 163 *See also* Bedrock;
        Soil, rocky
Rotation, 44 *See also* Development class;
        Forest management; Silvicultural
        measure; Wood production

S
Salix caprea See Goat willow
Salix pentandra See Bay willow
Sample plot, **25**, 64, 65
  angle count, 8, 23, **25**, 41, 50, 55, 73, 74
        (*see also* Tally tree)
  area represented by, 70, 72, 74
  centre of, **25**, 26, 70, 72, 86
  cluster of (*see* Cluster)

Index 267

dead wood, 27, **56**, 85
divided, 26
fixed radius, 24, 26, 55, 74, 85, 171
key habitat, 26, **47**, 71
keystone species, 55
locating, 27–28
maximum radius, **25**, 33, 72
number of, **18–19**, 70, 73, 85, 86, 87, 183
permanent (*see* Permanent plot)
Sample tree, **25**, 26, **50**, 75, 78, 82
 See also Tally tree; Tree
 measurements on, **51**, 59
Sampling *See also* Sample plot; Sampling design
 simulation of, **24–25**
Sampling density region, **18–19**, 21, 65, 69, 72, 87
Sampling design, **21**, 22, **25**, 87, 181, 183
 double sampling, 8, 23, 69, 180
 inclusion probability, 70, 72, 73
 simple random sampling, 88, 89
 stratification, 21, 23
 systematic, 23
Sampling error, 55, 69, 87, 88
 aggregates of calculation regions, **88–89**
 area estimates, **87**, 89, 181
 assessment of, 5, **88–89**, 180
 mean volume, **87**, 181
 simulated, **20**, 22
 total volume, **88–89**, 89, 182
Sand *See* Soil, sandy
Sand field, **48**, 256, 257, 259
Satellite image, 8, 20
Saw timber *See* Timber assortment
Scoliciosporum clorococcum, 55
Scots pine *See* Pine
Sea *See* Shore; Water
Seedling, 37, 38, 39, 40 *See also* Development class; Regeneration
 number of, 233
Seed tree, 37, 38 *See also* Development class
Shelter tree, 37, 38, **39**
 See also Development class
Shore, 36
Shrub, 30, **50**
Silver birch, **52, 57** *See also* Birch; Broad-leaved species
 diameter distribution, 223
 growing stock of, 89
 mean volume, 223
 stem number, 223
 volume by diameter class, 224

Silvicultural measure, 38, 48, 69, 84
 accomplished, **45–46**, 159–160, 238
 (*see also* Key habitat)
 proposed, **45–46**, 159–160
Silvicultural quality, 40, 44, **44–45**, 152, 156–157
 capable of further development, **44**, 156, 231
 by development class, 154–155, 231
 good, **44**, 154
 low-yielding, 38, **44**, 155, 231
 passable, **45**, 156
 reasons for decrease, **45**, 155, 156
 satisfactory, **45**, 156
Simulated error *See* Sampling error
Site class, **31**, **32**, 82, 88, 109–111, 147
 See also Mineral soil; Peatland
Site fertility, **32**, 33, 35, 46, 54, 78, 82, 88, 102, 103, 104, 118, 149
Small-leaved lime, **52**, 56, 222, 259, 260
Snag, 36, 37, 57, 172, 250, 251
Snow, damage caused by, 43, 153
Soil, 11, **33–35**
 grain size, **34**, 192
 mineral (*see* Mineral soil)
 mineral layer, 34
 organic, **34**, 101, 192
 organic layer, 34, **35**, 111
 peatland (*see* Peatland)
 preparation, 45, 160, 239, 240
 rocky, 32
 sandy, 32
 sorted, **34**, 101
 stoniness, 34, **34**, 36, 59, 101, 192, 193
 thickness of, 34
 type of, **34**, 101
Soil probe, 34
Sorbus aucuparia See European mountain ash
Spatial trend, 5, 88
Species *See* Keystone species; Meadow species; Threatened species; Tree species
Sphagnum, 33
Sphagnum fuscum, 32, 33, 48, 163
Spring, 37, 47, **48**, 255, 257, 258
Spruce, **52, 57**
 damage, 154
 dead, 168, 169, 170, 249, 250, 252
 diameter distribution, 223
 dominant species, 198, 199, 202–206
 as dominant species, 117, 131 (*see also* Peatland, spruce dominated)
 drain, 84

Spruce, (cont.)
  growing stock of, 89, 132
  increment of, 80, 82, 145
  mean volume, 126, 128, 129, 213–218
  mean volume, by development class, 206–211
  stem number, 222
  volume by diameter class, 224
  volume function, 75, **75–77**
Spruce mire *See* Peatland, spruce dominated
Stand, **26**
  boundaries, 6, 73 (*see also* Bi-stand; Centre point stand; Sample plot, divided)
  establishment type, **38**, 78, 146, 232 (*see also* Tree, origin type of)
Standard *See* Tree storey
Stem deformation, 42, 153
Stem number, 24, 36, 37, 38, 39, 40, **40**, 45, 69, 71, 72, **74**, 80, 82, 222, 259, 260
  dead trees, **58**
  saw timber, 136, 255–256
Stem volume *See* Volume
Stonefield, **48**
Stoniness *See* Soil, stoniness
Stratification *See* Sampling design
Stream, 37
Strip survey, 6
Stump, 8, 36, 37, **50**, 52, 57, **75**, 80, 250, 251
  class of, **53**
Summit, 32
Survivor tree *See* Tree, surviving
Swamp *See* Peatland
Swiss stone pine, **52**, 197
Systematic sampling *See* Sampling design

T
Tally tree, **25**, **49–55**, 71, 72, 73, 78, 80, 87
  See also Sample plot, angle count; Sample tree; Tree
  measurements on, **51**, 59
Talmeter tape, **59**
Taper curve, 54, 75, 77, 85
Tapio, 5
Taxation, 9, 35–36, **36**, 78
Tax Reform Committee, 6
Taxus baccata See English yew
Technical quality, 45, 46, 155 *See also* Timber assortment
Temperature *See also* Climatic variation
  effective temperature sum, 2–4, 24, **31**, 46, 78, 101, 137

Thinning *See* Cutting
Threatened species, 29, 46
Thuja, **52**
Tilia cordata See Small-leaved lime
Till *See* Glacial till
Timber assortment, 36, 51, 54, **55**, 75
  pulpwood, 37, 40, 51, **55**, 78, 219–221
  saw timber, 37, 40, 51, **55**, 78, 219–221
  volume by (*see* Volume)
  waste wood, 55
Timber production *See* Wood production
Timber production program, 9
Timber resources, 9
Time consumption, 20, 24
Time series, 63, 133, 138, 143
Topography, **33**
Training *See* Field crew
Transformed site, 35, **35**, 36, 47, 108, 195–196 *See also* Drainage
Transforming site, 35, **35**, 36, 47, 108, 195–196 *See also* Drainage
Tree *See also* Retention tree; Sample tree; Seed tree; Shelter tree; Tally tree
  age of (*see* Age)
  bark, 51, **53**, **57**, 59, 75, **80**, 81, 138
  branches, 41, 50, 53, 54, 55, 57, 75
  broken, 50, 51, 53, 57, 58
  bucking of, **54**, 136
  class of, 51, **51**, 78
  coordinates of, **50**, 51, 60
  crown height of, 51, **53**
  crown layer of, **53** (*see also* Tree storey)
  cut point of (*see* Cut point)
  dead (*see* Dead wood)
  definition of, **49**
  deformation of (*see* Stem deformation)
  diameter of (*see* Diameter)
  forked (*see* Fork tree)
  form factor, 75, **75**, 76, 79
  form height, **73**, 76, 78, 81
  growth of (*see* Diameter increment; Growth; Height increment; Volume increment)
  height of (*see* Height)
  increment of (*see* Diameter increment; Height increment; Volume increment)
  leaning, 42
  length of (*see* Height; Length, dead wood)
  living, **50**
  measurements on (*see* Sample tree; Tally tree)
  origin point of, **50**, 53

Index

origin type of, 51, **53**
small, **77**, 82
species of (*see* Tree species)
surviving, 80, **82–83**
type of, 51, **52**
vertical position of (*see* Tree, crown layer of)
volume of (*see* Volume)
Treeless *See* Development class; Regeneration, temporarily unstocked area
Tree ring, 60, 64, 140, 183
Tree species, 32, 37, 38, 39, 51, **52**, 69, 75, 78, 80, 82, 145
  all, **56**
  broad-leaved, 51, 52, 55, 57, 75, 80
  coniferous, 51, 52, 57, 75, 79, 80
  dead wood, **57**, 85, 168, 169, 170
  dominance, **38**, 40, 116–119, 121–124, 131, 212
  mixture, 119–120
  number of, 27
  proportions of, **38**
  rare, 55
  unfavorable, 155
Tree storey, 26, 36, **37**, 38, 41, 172
  See also Tree, crown layer of
  dominant, **38**, 39
  position of, **38**, 39
Trophic level *See* Site fertility
Two-phase sampling *See* Sampling design, double sampling

U

Ulmus glabra See Wych elm
Ulmus laevis See Fluttering elm
Ultrasound transponder, 59
Uncertainty *See* Bias; Measurement, accuracy of; Quality assurance; Sampling error
Under storey *See* Tree storey
Undrained peatland *See* Drainage
Uneven-aged *See* Age, uneven
United Nations Framework Convention on Climate Change (UNFCCC), 9, 10
Unproductive land, 24, **30**, 33, 35, 36, 50, 71 *See also* Land use
  area of, 186
  sampling error in area, 89
Upper diameter *See* Diameter
Upper storey *See* Tree storey
Usnea, 55

V

Vegetation zone, 3–4, 32, 48, 162
Vertex, 53, 59, **59**, 60, 64, 182
Volume, 20, 23, 24, 37 *See also* Dead wood; Keystone species
  by age class, 122, 130–131, 146, 202–205
  by development class, 130–132, 206–211
  by diameter class, 134–135, 224
  estimation of, 69, **78–79**
  on FAO forest and other wooded land, 202–205
  growing stock, 22, 46, 69, 71, 72, **72**, 86, 88, 132–134, 179, 213–218
  mean, 71, 72, **72, 73**, 89, 127, 129, 146, 213
  ocular estimation of, 5, 6
  by ownership, 132
  in protected areas, 149
  retention trees, 36
  sample tree, **77–78**, 79, 80, 81
  sampling error, 87, **87–89**, 89, 181
  saw timber, 135–136, 219–221
  by site class, 214
  tally tree, 72, **78–79**
  by timber assortment, **77–78**, 79
  time series, 133, 134
  total (*see* Volume, growing stock)
  tree, 51, 53, 59, 71, 75
Volume function, 75, 130, 180
  small trees, **77**
Volume increment, 69, 136
  by age class, 146
  be cite class, 227–229
  drain (*see* Drain, increment of)
  estimation of, 69, **85**
  growing stock, 80, 140, 145, 179
  mean, 30, 36, 137, 146
  sample tree, 80, **82**
  sampling error, 138
  by site class, 227
  small trees, 82
  survivor trees, **82–83**
  time series, 143
  total (*see* Volume increment, growing stock)
  tree, **80**

W

Water, 16, 23, 24, 25, **30**, 33, 35, 52
  See also Land use
Wilderness reserve, 29, 149

Willow, 33, 36, 50, 57 *See also* Bay willow; Goat willow
Wind, damage caused by, 43, 153
Wood production, 33, 35, 39, 46, 85, 121, 124, 152, 155, 159 *See also* Cutting; Forest management; Silvicultural measure
   growing stock on land available for, 132, 217–218
   implications of protection, 149
   land available for, 101, 189
   restrictions on, 28, 29, 47, 100, 170, 189
   sustainability of, 46
   taxation of (*see* Taxation)
World's Forestry Congress, 6
Wych elm, **52**, 56
   as keystone species, 259, 260

Y
Yard, 30, 50
Yew *See* English yew
Yield, 44, 53 *See also* Silvicultural quality; Site fertility; Wood production